TEXTS AND READINGS
IN MATHEMATICS **15**

Spectral Theory of Dynamical Systems

Second Edition

Spectral Theory of Dynamical Systems

Second Edition

M. G. Nadkarni

HINDUSTAN
BOOK AGENCY

Published by

Hindustan Book Agency (India)
P 19 Green Park Extension
New Delhi 110 016
India

Email: info@hindbook.com
www.hindbook.com

ISBN 978-93-86279-81-1

Contents

Preface

This book treats some basic topics in the spectral theory of dynamical systems, where by a dynamical system we mean a measure space on which a group of automorphisms acts preserving the sets of measure zero. The treatment is at a general level, but even here, two theorems which are not on the surface, one due to H. Helson and W. Parry and the other due to B. Host are presented. Moreover non-singular automorphisms are considered and systems of imprimitivity are discussed. Riesz products, suitably generalised, are considered and they are used to describe the spectral types and eigenvalues of rank one automorphisms. On the other hand topics such as spectral characterisations of various mixing conditions, which can be found in most texts on ergodic theory, and also the spectral theory of Gauss Dynamical Systems, which is very well presented in Cornfeld, Fomin, and Sinai's book on Ergodic Theory, are not treated in this book. A number of discussions and correspondence on email with El Abdalaoui El Houcein made possible the presentation of mixing rank one construction of D. S. Ornstein.

I am deeply indebted to G. R. Goodson. He has edited the book and suggested a number of corrections and improvements in both content and language.

M. G. Nadkarni

Preface to the Second Edition

The major change in this edition is that a new chapter titled Calculus of Generalized Riesz Products has been added. This is based on some recent work of the author with El Houcein El Abdalaoui and supplements the material presented elsewhere in the book. I am thankful to Hindustan Book Agency for their enthusiastic suggestion to bring out a second edition of this book that first appeared more than twenty years ago. The original source files of the manuscript were located with the help and cooperation of the production department of Birkhauser Verlag. It is a pleasure to record my thanks to Dr Thomas Hempfling who made it possible.

M. G. Nadkarni

Chapter 1

The Hahn-Hellinger Theorem

Definitions and the Problem

1.1. Let $\mathcal{H}$ be a complex separable Hilbert space, $\mathcal{E}$ the collection of orthogonal projections in $\mathcal{H}$, and $(X, \mathcal{B})$ a Borel space. A function $E : \mathcal{B} \to \mathcal{E}$ is called a spectral measure if $E(X) = I$ and $E(\bigcup_{i=1}^{\infty} A_i) = \sum_{i=1}^{\infty} E(A_i)$, for any pairwise disjoint collection $A_1, A_2, A_3, \ldots$, of sets in $\mathcal{B}$.

1.2. The equality $E(\bigcup_{i=1}^{\infty} A_i) = \sum_{i=1}^{\infty} E(A_i)$ has to be interpreted in the sense that for each $x \in \mathcal{H}$, $E(\bigcup_{i=1}^{\infty} A_i)x = \sum_{i=1}^{\infty} E(A_i)x$. We speak of E as being a spectral measure on $\mathcal{B}$ or on $\mathcal{H}$ depending on convenience.

1.3 Definition. Let E_1 be a spectral measure on a Hilbert space $\mathcal{H}_1$, E_2 a spectral measure on a Hilbert space $\mathcal{H}_2$, and let $(X, \mathcal{B})$ remain the same for E_1 and E_2. We say that E_1 and E_2 are unitarily equivalent if there exists an invertible isometry $S : \mathcal{H}_1 \leftrightarrow \mathcal{H}_2$ such that for all $A \in \mathcal{B}$,

$$SE_1(A)S^{-1} = E_2(A).$$

1.4. When are two spectral measures on $(X, \mathcal{B})$ unitarily equivalent? An answer to this question is provided by the Hahn-Hellinger theorem which we prove in this chapter. We begin with the simple case of this theorem when E_1 and E_2 are multiplication by indicator functions in L^2 of two measures μ and ν respectively (see **1.6**).

1.5. Let μ be a finite or a σ-finite measure on $(X, \mathcal{B})$ such that $L^2(X, \mathcal{B}, \mu)$ is separable. If $B \in \mathcal{B}$, then $E(B)$ defined by $E(B)f = 1_B f, f \in L^2(X, \mathcal{B}, \mu)$ is an orthogonal projection on the subspace of functions in $L^2(X, \mathcal{B}, \mu)$ vanishing outside B, and, $B \to E(B)$ is a spectral measure which we denote by E_μ. Suppose we have another finite or σ-finite measure ν on $(X, \mathcal{B})$ such that $L^2(X, \mathcal{B}, \nu)$ is separable. When are E_μ and E_ν unitarily equivalent? The answer is given by:

1.6 Proposition. E_μ and E_ν are unitarily equivalent if and only if μ and ν have the same null sets.

Proof. Suppose μ and ν have the same null sets. We then set up an invertible isometry S between $L^2(X, \mathcal{B}, \mu)$ and $L^2(X, \mathcal{B}, \nu)$ as follows:

$$Sf = \sqrt{\frac{d\mu}{d\nu}} f, f \in L^2(X, \mathcal{B}, \mu).$$

For all $B \in \mathcal{B}, f \in L^2(X, \mathcal{B}, \nu)$,

$$
\begin{aligned}
SE_\mu(B)S^{-1}f &= SE_\mu(B)\sqrt{\frac{d\nu}{d\mu}} f \\
&= S1_B\sqrt{\frac{d\nu}{d\mu}} f \\
&= 1_B f\sqrt{\frac{d\nu}{d\mu}}\sqrt{\frac{d\mu}{d\nu}} \\
&= 1_B f = E_\nu(B)f.
\end{aligned}
$$

Thus E_μ and E_ν are unitarily equivalent. Conversely if E_μ and E_ν are unitarily equivalent, then there exists an isometry $S : L^2(X, \mathcal{B}, \mu)$ onto $L^2(X, \mathcal{B}, \nu)$ such that

$$SE_\mu(B)S^{-1} = E_\nu(B)$$

for all $B \in \mathcal{B}$. Thus

$$\mu(B) = 0 \Leftrightarrow E_\mu(B) = 0 \Leftrightarrow E_\nu(B) = 0 \Leftrightarrow \nu(B) = 0.$$

This shows that μ and ν have the same null sets.

The Case of Multiplicity One, Cyclic Vector

1.7 Definition. A spectral measure E is said to be of simple multiplicity or of multiplicity one if there exists an $x \in \mathcal{H}$ such that $\mathcal{H}$ is the closed linear span of $\{E(A)x : A \in \mathcal{B}\}$. Any such vector x is then called a cyclic vector for E.

1.8. Suppose E is a spectral measure and $x \in \mathcal{H}$. We write $\mathcal{H}_x$ to denote the closed linear span of $\{E(A)x : A \in \mathcal{B}\}$. Then for all $A \in \mathcal{B}$ and $y \in \mathcal{H}_x, E(A)y \in \mathcal{H}_x$. We can therefore restrict E to $\mathcal{H}_x$. The restriction of E to $\mathcal{H}_x$, denoted by E_x, is of multiplicity one; x being a cyclic vector for E_x

1.9. The spectral measure E_μ on $L^2(X, \mathcal{B}, \mu)$ is always of multiplicity one. If $\mu(X) < \infty$, then $\{E_\mu(B)1 : B \in \mathcal{B}\} = \{1_B : B \in \mathcal{B}\}$ spans $L^2(X, \mathcal{B}, \mu)$. If μ is infinite but σ-finite, then one can find an $f \in L^2(X, \mathcal{B}, \mu)$ such that $f > 0$ a.e. The collection $\{E_\mu(B)f : B \in \mathcal{B}\}$ then spans $L^2(X, \mathcal{B}, \mu)$.

1.10. Discussions of this section will yield a proof of the Hahn-Hellinger theorem for spectral measures of multiplicity one. Let E be a spectral measure on a Hilbert space $\mathcal{H}$ and let $x \in \mathcal{H}$. Let $\mu_x(B) = (E(B)x, x)$. Then

$$\mu_x(B) = (E(B)x, E(B)x) \geq 0, \mu_x(X) = (E(X)x, x) = (x, x) = \|x\|^2.$$

Further countable additivity of E implies that of μ_x. Thus μ_x is a finite non-negative countably additive measure on $\mathcal{B}$. Further

$$\| E(B)x \|^2 = \mu_x(B) = \int_X | 1_B(x) |^2 \, d\mu_x.$$

If $B_1, B_2, \ldots, B_n$ are pairwise disjoint sets in $\mathcal{B}$ and $c_1, c_2, \ldots, c_n$ are complex numbers, then

$$\left\| \sum_{i=1}^n c_i E(B_i)x \right\|^2 = \sum_{i=1}^n | c_i |^2 \, (E(B_i)x, x)$$

$$= \sum_{i=1}^n | c_i |^2 \, \mu_x(B_i) = \int_X \left| \sum_{i=1}^n c_i 1_{B_i} \right|^2 \, d\mu_x.$$

More generally, for pairwise disjoint sets $A_1, A_2, \ldots, A_m$ and pairwise disjoint sets $B_1, B_2, \ldots, B_n$ in $\mathcal{B}$ we have

$$\left(\sum_{i=1}^m c_i E(A_i)x, \sum_{j=1}^n d_j E(B_j)x \right) = \sum_{i=1}^m \sum_{j=1}^n c_i \bar{d}_j \mu_x(A_i \cap B_j)$$

$$= \int_X \sum_{i=1}^m c_i 1_{A_i} \sum_{j=1}^n \bar{d}_j 1_{B_j} \, d\mu_x.$$

We can therefore define an invertible isometry S between the preHilbert space of all finite linear combinations $\sum_{i=1}^n c_i E(B_i)x$ and the preHilbert space of simple functions of the form $\sum_{i=1}^n c_i 1_{B_i}$,

$$S \left(\sum_{i=1}^n c_i E(B_i)x \right) = \sum_{i=1}^n c_i 1_{B_i}.$$

The operator S extends to an invertible isometry from all of $\mathcal{H}_x$ onto $L^2(X, \mathcal{B}, \mu_x)$. Further

$$SE(A)S^{-1}1_B = SE(A)E(B)x = SE(A \cap B)x = 1_{A \cap B} = 1_A \cdot 1_B.$$

This shows that if E_x is the restriction of E to $\mathcal{H}_x$ then $SE_x(A)S^{-1} = E_{\mu_x}(A)$ for all $A \in \mathcal{B}$, i.e., E_{μ_x} and E_x are unitarily equivalent.

Suppose now that multiplicity of E is one and $x \in \mathcal{H}$ is a cyclic vector for E. Then $E_x = E$ and E is unitarily equivalent to E_{μ_x}. If y is another cyclic vector for E then E is unitarily equivalent to E_{μ_y} acting on $L^2(X, \mathcal{B}, \mu_y)$. Thus E_{μ_x} and E_{μ_y} are unitarily equivalent. By proposition **1.6** μ_x and μ_y have the same null sets. Thus we have proved:

1.11 Theorem. *Let E be a spectral measure on $(X, \mathcal{B})$ of multiplicity one. Then there exists a finite non-negative measure μ on $(X, \mathcal{B})$ and an invertible isometry S from $\mathcal{H}$ (on which E is defined) onto $L^2(X, \mathcal{B}, \mu)$ such that for all $f \in L^2(X, \mathcal{B}, \mu)$,*

$$SE(B)S^{-1}f = 1_B f, \quad B \in \mathcal{B},$$

i.e., E and E_μ are unitarily equivalent. If ν is another such measure then μ and ν have the same null sets.

1.12. If μ is a σ-finite measure on $\mathcal{B}$ then the collection all σ-finite measures on $\mathcal{B}$ having the same null sets as μ is called the measure class of μ. If E is a spectral measure of multiplicity one, x and y are cyclic vectors for E, then μ_x and μ_y are in the same measure class. If z is another vector in $\mathcal{H}$, then μ_z is absolutely continuous with respect to μ_x for any cyclic vector x.

1.13. If E is a spectral measure of multiplicity one, then associated with it is the measure class of μ_x, where x is a cyclic vector for E, and E and E_{μ_x} are unitarily equivalent. If F is another spectral measure of multiplicity one on $(X, \mathcal{B})$, then we can associate with F the measure class of the measure ν_y, where y is a cyclic vector for F. Further F is unitarily equivalent to E_{ν_y}. Finally by proposition **1.6** we see that E and F are unitarily equivalent if and only if μ_x and ν_y belong to the same measure class.

1.14. For a spectral measure E of multiplicity one the measure class of μ_x, where x is a cyclic vector, is called the maximal spectral type of E. Two spectral measures of multiplicity one with the same maximal spectral type are unitarily equivalent. The maximal spectral type of a spectral measure of multiplicity one is therefore a complete invariant of unitary equivalence for such spectral measures.

Exercise. Let E be a spectral measure and let $x \in \mathcal{H}$. Let ν be a finite measure absolutely continuous with respect to μ_x. Show that there exists $y \in \mathcal{H}_x$ such that $\mu_y = \nu$.

Application to Second Order Stochastic Processes

1.15. We now digress to give an application of the above considerations to the representation of second order stochastic processes. These applications, in the generality given here, are due to Cramér (See [1]). Let $(\Omega, \mathcal{A}, P)$ be a probability space and assume that $L^2(\Omega, \mathcal{A}, P)$ is separable. A function from the real line $\mathbb{R}$ into $L^2(\Omega, \mathcal{A}, P)$ is called a second order stochastic process. We denote the value of the function at t by X_t, and, the process by $(X_t)_{t\in\mathbb{R}}$. We assume that $E(X_t) = 0$ for all t, where $E(X_t) = \int X_t dP =$ expected value of X_t. Let $\mathcal{M}_t =$ closed subspace of $L^2(\Omega, \mathcal{A}, P)$ spanned by $\{X_s : s \le t\}$. If $s < t$, then $\mathcal{M}_s \subseteq \mathcal{M}_t$. We write $\mathcal{M}_\infty$ for the closure of $\bigcup_{s<\infty} \mathcal{M}_s$ and $\mathcal{M}_{-\infty}$ for the intersection $\bigcap_{s<\infty} \mathcal{M}_t$. The process $(X_t)_{t\in\mathbb{R}}$ is called deterministic if $\mathcal{M}_{-\infty} = \mathcal{M}_\infty$, non-deterministic otherwise. It is called purely non-deterministic if $\mathcal{M}_{-\infty} = \{0\}$ but $\mathcal{M}_\infty \ne \{0\}$. For any t, where $t \in \mathbb{R}$ or $t = -\infty$ or $t = \infty$, let P_t denote the orthogonal projection on the subspace $\mathcal{M}_t$. Let $Y_t = P_{-\infty}X_t$ and $Z_t = X_t - Y_t$. Then

$$Z_t \in \mathcal{M}_t \ominus \mathcal{M}_{-\infty} \underset{\text{def}}{=} \mathcal{N}_t.$$

Further $X_t = Y_t + Z_t, t \in \mathbb{R}$. If we write $\mathcal{M}_t(X)$ to denote the closed subspace spanned by $\{X_s : s \le t\}$, then we see that

$$\mathcal{M}_t(X) \subseteq \mathcal{M}_t(Y) \oplus \mathcal{M}_t(Z) \subseteq \mathcal{M}_{-\infty}(X) \oplus \mathcal{N}_t(X) = \mathcal{M}_t(X)$$

where

$$\mathcal{N}_t(X) = \mathcal{M}_t(X) \ominus \mathcal{M}_{-\infty}(X).$$

Further $\mathcal{M}_t(Y) \subseteq \mathcal{M}_{-\infty}(X)$ and $\mathcal{M}_t(Z) \subseteq \mathcal{N}_t(X)$. Thus $\mathcal{M}_t(Y) = \mathcal{M}_{-\infty}(X)$ and $\mathcal{M}_t(Z) = \mathcal{N}_t(X)$. This show that $(Y_t)_{t\in\mathbb{R}}$ is a deterministic process and $(Z_t)_{t\in\mathbb{R}}$ is purely non-deterministic. Thus every second order stochastic process is a sum of two processes, one deterministic and the other purely non-deterministic.

1.16. Now assume that $(X_t)_{t\in\mathbb{R}}$ is purely non-deterministic. Define $Q_t = \lim_{s\to t+} P_s$, where the limit is taken from the right and in the sense that for each x, $Q_t x = \lim_{s\to t+} P_s x$. We define $E(a, b] = Q_b - Q_a$; E gives rise to a spectral measure on the Borel subsets of $\mathbb{R}$. We assume that E has multiplicity one. Let Z be a cyclic vector for E. Let $Q_t Z = Z_t$. Then $Z_t - Z_s$ and $Z_b - Z_a$ are orthogonal whenever $[s, t] \cap [a, b]$ is empty. Thus $(Z_t)_{t\in\mathbb{R}}$ is a process with orthogonal increments. The collection $\{Z_t - Z_s : -\infty < s < t < \infty\}$ further spans $\mathcal{M}_\infty$. Define $m(a, b] = \|Z_b - Z_a\|^2, a < b$; m extends to a countably additive measure on $\mathbb{R}$ which is indeed the maximal spectral type of E. There is an isometry S between $\mathcal{M}_\infty$ and $L^2(\mathbb{R}, \mathcal{B}_\mathbb{R}, m)$ such that $S(Z_b - Z_a) = 1_{(a,b]}$, and for all $B \in \mathcal{B}_\mathbb{R}$,

$$SE(B)S^{-1}f = 1_B f, \quad f \in L^2(\mathbb{R}, \mathcal{B}_\mathbb{R}, m).$$

Now $X_t \in Q_t \mathcal{M}_\infty$. Therefore X_t is a limit of finite linear combinations of the form $\sum_{j=1}^{n} c_j(Z_{t_j} - Z_{t_{j-1}})$ where $t_1, t_2, \ldots, t_n$ are all less than or equal to t. The corresponding functions $\sum_{j=1}^{n} c_j 1_{(t_{j-1}, t_j]}$ then converge in $L^2(\mathbb{R}, \mathcal{B}_\mathbb{R}, m)$ to a function f_t which is indeed SX_t. f_t vanishes for $s > t$. We codify this information by writing X_t as a stochastic integral:

$$X_t = \int_{-\infty}^{t} f_t(s)dZ_s,$$

where $(Z_t)_{t\in\mathbb{R}}$ is the process with orthogonal increments constructed above. Thus we have proved:

1.17 Theorem. *Every purely non-deterministic second order stochastic process* $(X_t)_{t\in\mathbb{R}}$ *,* $E(X_t) = 0$*, such that the associated resolution of identity* $Q_t, t \in \mathbb{R}$*, on* $\mathcal{M}_\infty$ *has multiplicity one, can be represented in the form:*

$$X_t = \int_{-\infty}^{t} f_t(s)dZ_s,$$

where $(Z_s)_{s\in\mathbb{R}}$ *is a process with orthogonal increments such that for each t,* $Q_t\mathcal{M}_\infty = \bigcap_{s>t} \mathcal{M}_s$ *is same as the subspace spanned by*

$$\{Z_b - Z_a : a < b,\ a,\ b \le t\}.$$

1.18. It should be remarked that the requirement on the spectral measure E (given by the resolution of identity $Q_t, t \in \mathbb{R}$) to have multiplicity one can not be removed since there are scalar second order stochastic processes (with expectation zero) for which E has multiplicity bigger than one. Although some simpler examples may be given to illustrate this fact, we mention the following example which may be of some interest.

1.19. Let $(X_t)_{t\in\mathbb{R}}, (Y_t)_{t\in\mathbb{R}}$ be two second order stochastic processes such that:

(i) $E(X_t) = 0, E(Y_t) = 0$ for all t,
(ii) they are purely non-deterministic,
(iii) they are mean continuous, i.e., for all s

$$E\mid X_t - X_s \mid^2,\ E\mid Y_t - Y_s \mid^2 \to 0\ \ as\ \ t \to s,$$

(iv) they are second order stationary, i.e., for all t and s

$$E(X_t \cdot \overline{X}_s) = E(X_{t-s} \cdot \overline{X}_0),\quad E(Y_t \cdot \overline{Y}_s) = E(Y_{t-s} \cdot \overline{Y}_0),$$

(v) for all s, t, $X_t \perp Y_s$ (it is assumed that the processes are defined on the same probability space).

Now it is known that under these conditions both $(X_t)_{t\in\mathbb{R}}$, $(Y_t)_{t\in\mathbb{R}}$ have the associated spectral measures, E and F say, of simple multiplicity. Further, the process $Z_t = X_t + Y_t, t \in \mathbb{R}$, is purely non-deterministic and of multiplicity one in the sense that the associated spectral measure is of multiplicity one. However, if we consider $W_t = X_t + C(t)Y_t$, where $C(t)$ is a non-measurable character on $\mathbb{R}$, then $(W_t)_{t\in\mathbb{R}}$ is purely non-deterministic and second order stationary (but not mean continuous) with an associated spectral measure of multiplicity two since the spectral measure of the discontinuous (in the usual topology of $\mathbb{R}$) unitary group of operators associated with the process $W_t, t \in \mathbb{R}$, is supported on the natural embedding of the real line in the Bohr group and its translate by the discontinuous character C. (See **12.14 (iii)**.)

1.20. The above method can be used to give an example of a second order stationary purely non-deterministic (not mean continuous, however) stochastic process such that the associated spectral measure has any desired multiplicity, finite or infinite.

Spectral Measures of Higher Multiplicity: A Canonical Example

1.21. We now return to the consideration of spectral measures which are not necessarily of multiplicity one. We first give some more examples of spectral measures which are not of multiplicity one.

1.22 Example. Let μ be a finite measure on $(X, \mathcal{B})$; $\mathcal{H} = L^2(X, \mathcal{B}, \mu) \oplus L^2(X, \mathcal{B}, \mu)$. We assume that $L^2(X, \mathcal{B}, \mu)$ is separable. Define E by $E(B)f = 1_B f$, where $f = (f_1, f_2) \in \mathcal{H}, B \in \mathcal{B}$. If μ is not the null measure then E does not have multiplicity one. For if E has multiplicity one then there exists $f = (f_1, f_2)$ such that $\{1_B f : B \in \mathcal{B}\}$ spans $\mathcal{H}$ so that if $g \in \mathcal{H}$, then for a.e. $x, g(x) = c(x)f(x)$, where c is some measurable complex valued function on X. Therefore the functions $e_1 : e_1(x) = (1, 0)$ *and* $e_2 : e_2(x) = (0, 1)$ are expressible as $c_1(x)f(x)$ and $c_2(x)f(x)$ respectively for some complex valued measurable functions c_1 and c_2, which is clearly impossible.

1.23 Example. Let $\mu_1, \mu_2, \mu_3, \ldots$ be a sequence of non-trivial finite measures on $(X, \mathcal{B})$ such that each $L^2(X, \mathcal{B}, \mu_n)$ is separable. Let $\mathcal{H} = \sum_{n=1}^{\infty} L^2(X, \mathcal{B}, \mu_n)$. Define E_μ by $E_\mu(B)f = 1_B f = (1_B f_1, 1_B f_2, \ldots)$, where $B \in \mathcal{B}$, and $f = (f_1, f_2, f_3, \ldots) \in \mathcal{H}$. If the μ_i's are mutually singular, then E_μ has multiplicity one since $\mathcal{H}$ is then indeed equal to $L^2(X, \mathcal{B}, \sum_{n=1}^{\infty} \mu_n)$. But as soon as two of the μ_i's, say μ_1 and μ_2, are not mutually singular, E_μ does not have multiplicity one.

1.24. We write $\mu_1 \gg \mu_2$ if μ_2 is absolutely continuous with respect to μ_1. Assume that in **1.23** $\mu_1 \gg \mu_2 \gg \mu_3 \gg \cdots$. Then E_μ does not have multiplicity one unless μ_2 is the null measure. This example is canonical in the sense that any spectral measure E on $(X, \mathcal{B})$ is unitarily equivalent to an E_μ for some suitable sequence $\mu_1 \gg \mu_2 \gg \mu_3 \gg \cdots$, of finite measures. The measure classes of $\mu_i, i = 1, 2, 3, \ldots$, are uniquely determined by E. This is the content of a version of Hahn-Hellinger theorem which we proceed to prove.

Linear Operators Commuting with Multiplication

1.25 Lemma. *Let T be a bounded linear operator on $L^2(X, \mathcal{B}, \mu)$ such that for all $B \in \mathcal{B}$ and $f \in L^2(X, \mathcal{B}, \mu)$, $T1_B f = 1_B T f$, i.e., T commutes with all $E_\mu(B), B \in \mathcal{B}$, then there exists a function $\varphi \in L^2(X, \mathcal{B}, \mu)$, unique up to μ null sets, such that for all $f \in L^2(X, \mathcal{B}, \mu)$, $T f = \varphi f$.*

Proof. First assume that $\mu(X) < \infty$. Write $T1 = \varphi$. We will show that T is multiplication by $\varphi : Tf = \varphi f$ for all $f \in L^2(X, \mathcal{B}, \mu)$. First note that for all $B \in \mathcal{B}$, $T1_B = 1_B T1 = 1_B \varphi = \varphi 1_B$. Hence on the class $\mathcal{M}$ of finite linear combinations of indicator functions the operator T agrees with T_φ defined by $T_\varphi f = \varphi f$. Since T is bounded, the operator T_φ is bounded too. T_φ therefore extends to a bounded linear operator on $L^2(X, \mathcal{B}, \mu)$; T_φ and T agree on $L^2(X, \mathcal{B}, m)$. It is easy to see that φ is an L^∞ function; the extended T_φ has the form $T_\varphi f = \varphi f$, $f \in L^2(X, \mathcal{B}, \mu)$. Thus $T = T_\varphi$ and $\|\varphi\|_\infty = \|T\|$. This proves the lemma when $\mu(X) < \infty$.

If $\mu(X) = \infty$ but μ is σ-finite, then there exist pairwise disjoint sets $A_1, A_2, A_3, \ldots$ in $\mathcal{B}$ with union X and such that $\mu(A_i) < \infty$ for all i. Since T commutes with $E_\mu(A_i)$, T leaves the subspaces $L^2(X, \mathcal{B}, \mu \mid_{A_i})$ invariant. The restriction of T to this subspace commutes with E_i, where E_i denotes the restriction of E_μ to $L^2(X, \mathcal{B}, \mu\mid_{A_i})$. By the above considerations there exist φ_i, bounded by the norm of T, such that φ_i vanishes outside A_i and $Tf = \varphi_i f$ whenever $f \in L^2(X, \mathcal{B}, \mu \mid_{A_i})$. We set

$$\varphi = \varphi_1 + \varphi_2 + \varphi_3 + \cdots .$$

Clearly $\varphi \in L^\infty(X, \mathcal{B}, \mu)$ and $T = T_\varphi$. Also φ is unique up to μ-null sets.

1.26. Let n be a cardinal number $\leq \aleph_0$ and let $nL^2(X, \mathcal{B}, \mu)$ denote the direct sum of n copies of $L^2(X, \mathcal{B}, \mu)$. We may also view $nL^2(X, \mathcal{B}, \mu)$ as $L^2(X, \mathcal{B}, \mu, K)$, the space of square integrable functions taking values in the Hilbert space K, where $K = \mathbb{C}^n$ if $n < \infty$ and $K = l^2$ if $n = \aleph_0$. If $n = \aleph_0$ then $nL^2(X, \mathcal{B}, \mu)$ is to be understood as the space of sequences $(f_1, f_2, f_3, \ldots)$ of functions in $L^2(X, \mathcal{B}, \mu)$ such that $\sum_{k=1}^{\infty} \| f_k \|^2 < \infty$, with inner product between $(f_1, f_2, f_3 \ldots)$ and $(g_1, g_2, g_3 \ldots)$ given by $\sum_{k=1}^{\infty} (f_k, g_k)$.

1.27 Lemma. *Let T be a bounded linear operator on $nL^2(X, \mathcal{B}, \mu)$ such that for all $f = (f_1, f_2, \ldots)$ in $nL^2(X, \mathcal{B}, \mu)$ and for all $B \in \mathcal{B}$, $T1_B f = 1_B Tf$. Then there exists a unique (up to μ-null sets) $n \times n$ matrix valued function $\tilde{T}$ with measurable entries such that for all $f \in nL^2(X, \mathcal{B}, \mu)$,*

$$(Tf)(x) = \tilde{T}(x) f(x) \quad a.e.$$

For a.e. x, $\tilde{T}(x)$ defines an operator on K whose norm is less than or equal to $\| T \|$.

Proof. Let $e_i = (0, \ldots, 0, 1, 0 \ldots, 0)$ where 1 occurs in i^{th} place; $e_i \in \mathbb{C}^n$ or $e_i \in l^2$ depending on whether $n < \aleph_0$ or $n = \aleph_0$. We think of e_i as an element in K as well as an element in $nL^2(X, \mathcal{B}, \mu)$.

Write $Te_i = \varphi_i = (\varphi_{i1}, \varphi_{i2}, \ldots) \in nL^2(X, \mathcal{B}, \mu)$ and let φ_i form, for each i, the ith column of a matrix $\tilde{T}$. Let $\mathcal{M}$ denote the dense linear manifold of $nL^2(X, \mathcal{B}, \mu)$ of functions of the form $\xi = (\xi_1, \xi_2, \ldots)$, where only finitely many ξ_i 's are non-zero, and, those that are non-zero are finite linear combinations of indicator functions of sets in $\mathcal{B}$. We define a linear operator on $\mathcal{M}$, also denoted by $\tilde{T}$, as follows: $\tilde{T}\xi = \sum_{i=1}^{\infty} \xi_i\varphi_i$. Now

$$\tilde{T}\xi = \sum_{i=1}^{\infty} \xi_i\varphi_i = \sum_{i=1}^{\infty} \xi_i Te_i = T\left(\sum_{i=1}^{\infty} \xi_i e_i\right),$$

where the last equality is valid because T commutes with multiplication by 1_B for all $B \in \mathcal{B}$. Thus $\tilde{T}$ and T agree on a dense linear manifold of $nL^2(X, \mathcal{B}, \mu)$, hence $\tilde{T}$ extends to a bounded linear operator on all of $nL^2(X, \mathcal{B}, \mu)$. This extension of $\tilde{T}$, which we again denote by $\tilde{T}$, agrees with T. Thus $Tf = \tilde{T}f$.

Let $t(x)$ denote the operator norm of $\tilde{T}(x)$ on K. Let $y_1, y_2, y_3, \ldots$, be a countable collection of vectors in K, each y_i of norm 1, such that the collection is dense in $\{y \in K : \| y \| = 1\}$. Now $t(x) = \sup \| \tilde{T}(x)y_i \|$, where the supremum is taken over $y_1, y_2, y_3 \ldots$. Thus t is a measurable function. We show that $t(x) \leq \| T \|$ a.e. If not, $t(x) > \| T \|$ on a set A of positive measure. We can find an element $\xi \in nL^2(X, \mathcal{B}, \mu)$, vanishing outside A, such that $\| \tilde{T}(x)\xi(x) \| > \| T \|$ and $\| \xi(x) \| = 1$ for $x \in A$. If we write $\psi(x) = (\mu(A))^{-1/2}\xi(x)$, then ψ has norm one in $nL^2(X, \mathcal{B}, \mu)$ and the squared norm of $T\psi$ is

$$\int_X \| (\mu(A)^{-1/2})\tilde{T}(x)\xi(x) \|^2 \, d\mu = \int_A \| \tilde{T}(x)\xi(x) \|^2 \frac{1}{\mu(A)} d\mu > \| T \|^2,$$

which is a contradiction. Hence $t(x) \leq \| T \|$ a.e. μ. It is clear that $\tilde{T}$ is unique up to a set of μ measure zero. The lemma is proved.

We restate this lemma in an alternative form.

1.28 Alternative form of Lemma 1.27. *Let K be a complex separable Hilbert space and let $\mathcal{H} = L^2(X, \mathcal{B}, \mu, K)$. Let T be a bounded linear operator on $\mathcal{H}$ such that for all $f \in \mathcal{H}$ and $B \in \mathcal{B}$,*

$$T1_B f = 1_B Tf.$$

Then there exists a weakly measurable function $\tilde{T}$ on X whose values are bounded linear operators on K such that $\| \tilde{T}(x) \| \leq \| T \|$ a.e. and $(Tf)(x) = \tilde{T}(x)f(x)$ a.e. where $f \in \mathcal{H}$. $\tilde{T}$ is unique up to a set of measure zero.

Remark 1. Weak measurability of $\tilde{T}$ means that for all $u, v \in K$, $(\tilde{T}(x)u, v)$ is a complex valued measurable function on X.

Remark 2. If T is as in **1.27** (or **1.28**) then T^* also commutes with multiplication by 1_B for all $B \in \mathcal{B}$. Hence T^* is of the form

$$(T^* f)(x) = \tilde{T}_1(x)f(x);$$

$\tilde{T}_1(x)$ is indeed equal to $\tilde{T}^*(x)$ for a.e. x. If T is unitary then so is $\tilde{T}^*(x)$ for a.e. x.

1.29. If E is a spectral measure on $\mathcal{H}$ and $\mathcal{H}_1 \subseteq \mathcal{H}$ is a subspace such that for all $B \in \mathcal{B}$, $E(B)\mathcal{H}_1 \subseteq \mathcal{H}_1$ then for all $B \in \mathcal{B}$, $E(B)\mathcal{H}_1^{\perp} \subseteq \mathcal{H}_1^{\perp}$, where $\mathcal{H}_1^{\perp}$ denotes the orthogonal complement of $\mathcal{H}_1$ in $\mathcal{H}$. To see this let $f \in \mathcal{H}_1^{\perp}$ and $g \in \mathcal{H}_1$. Now $(E(B)f, g) = (f, E(B)g) = 0$ since $E(B)g \in \mathcal{H}_1$. This holds for all $g \in \mathcal{H}_1$ so that $E(B)f \in \mathcal{H}_1^{\perp}$. Thus $E(B)\mathcal{H}_1^{\perp} \subseteq \mathcal{H}_1^{\perp}$. If P denotes the orthogonal projection on $\mathcal{H}_1$, then $PE(B) = E(B)P$ for all $B \in \mathcal{B}$. These observations are needed in the proof of:

1.30 Corollary to 1.27 or 1.28. *Let $\mathcal{H}_1 \subseteq L^2(X, \mathcal{B}, \mu, K) = \mathcal{H}$ be a closed subspace such that for all $B \in \mathcal{B}$, $E_\mu(B)\mathcal{H}_1 \subseteq \mathcal{H}_1$. Then there exists a weakly measurable function J on X (values of J being orthogonal projections in K) such that*

$$\mathcal{H}_1 = J\mathcal{H} \underset{\mathrm{def}}{=} \{f \in \mathcal{H} : f(x) \in J(x)K \quad a.e.\}.$$

Proof. Let P denote the orthogonal projection on $\mathcal{H}_1$. Then

$$PE_\mu(B) = E_\mu(B)P$$

for all $B \in \mathcal{B}$. Hence by **1.27** there exists a weakly measurable function $\tilde{P}$ whose values are bounded operators on K of norm ≤ 1 such that $(Pf)(x) = \tilde{P}(x)f(x)$ *a.e.* Now

$$\tilde{P}^2(x)f(x) = (P^2 f)(x) = (Pf)(x) = \tilde{P}(x)f(x) \quad a.e.$$

Thus $\tilde{P}(x)$ is a projection a.e. Further $\tilde{P}(x)$ is self adjoint for a.e. x since P is self adjoint. If we write J for $\tilde{P}$ we have the desired result.

1.31 Corollary to 1.28. *Let μ and ν be finite measures on $(X, \mathcal{B})$ and let K and L be complex separable Hilbert spaces of dimension k and l respectively. Let S be an isometry from $L^2(X, \mathcal{B}, \mu, K)$ onto $L^2(X, \mathcal{B}, \nu, L)$ such that for all $B \in \mathcal{B}$ and $f \in L^2(X, \mathcal{B}, \nu, L)$,*

$$S1_B S^{-1} f = 1_B f.$$

Then μ and ν have the same null sets, $k = l$, and there exists a weakly measurable function $\tilde{S}$ on X whose values are invertible isometries from K onto L such that for $f \in L^2(X, \mathcal{B}, \mu, K)$,

$$(Sf)(x) = \tilde{S}(x)f(x)\sqrt{\frac{d\mu}{d\nu}}(x) \quad a.e. \ \mu.$$

Proof. The spectral measure $E_\mu : E_\mu(B)f = 1_B f$ defined on $L^2(X, \mathcal{B}, \mu, K)$ is unitarily equivalent to E_ν defined similarly on $L^2(X, \mathcal{B}, \nu, L)$. Hence μ and ν are in the same measure class. Define T from $L^2(X, \mathcal{B}, \mu, K)$ onto $L^2(X, \mathcal{B}, \mu, L)$ by

$$Tf = Sf\sqrt{\frac{d\nu}{d\mu}}.$$

Then $T1_B f = 1_B Tf$ for all $B \in \mathcal{B}$ and $f \in L^2(X, \mathcal{B}, \mu, K)$. By **1.29** there exists a weakly measurable operator valued function $\tilde{S}$ (values of $\tilde{S}$ being bounded linear operators from K into L) such that

$$(Tf)(x) = \tilde{S}(x)f(x) \quad a.e. \ \mu.$$

Since T is an invertible isometry, we conclude that for a.e. x, $\tilde{S}(x)$ is an invertible isometry from K into L. Clearly then $k = l$. Now

$$(Tf)(x) = (Sf)(x)\sqrt{\frac{d\nu}{d\mu}}(x) = \tilde{S}(x)f(x) \quad a.e. \ \mu,$$

whence

$$(Sf)(x) = \tilde{S}(x)f(x)\sqrt{\frac{d\mu}{d\nu}}(x) \quad a.e.\ \mu.$$

The corollary is proved.

Spectral Type; Maximal Spectral Type

1.32. Let E be a spectral measure on $(X, \mathcal{B})$, $\mathcal{H}$ being the underlying Hilbert space. A finite or a σ-finite measure μ, or the measure class of such a measure is called a spectral type of E if there exists an $x \in \mathcal{H}$ such that $\mu_x : \mu_x(B) = (E(B)x, x)$, $B \in \mathcal{B}$ has the same null sets as μ. A finite or a σ-finite measure μ, or the measure class of such a measure is called the maximal spectral type of E if μ has the same null sets as E. We shall show in **1.32 (continued)** that the maximal spectral type exists by exhibiting a $z \in \mathcal{H}$ such that μ_z has the same null sets as E. Every other spectral type of E is absolutely continuous with respect to the maximal spectral type of E. If μ is a finite or a σ-finite measure on $\mathcal{B}$, then the subspace $\mathcal{H}_\mu = \{x \in \mathcal{H} : \mu \gg \mu_x\}$ is invariant under E and we can restrict E to $\mathcal{H}_\mu$. The restriction of E to $\mathcal{H}_\mu$ is defined by $E_1 : E_1(B) = P_{\mathcal{H}_\mu} E(B)$, $B \in \mathcal{B}$, where $P_{\mathcal{H}_\mu}$ denotes the orthogonal projection on $\mathcal{H}_\mu$. Suppose F is another spectral measure defined on $(X, \mathcal{B})$ acting in a complex separable Hilbert space $\mathcal{K}$. We can consider $\mathcal{K}_\mu$, the space of vectors $y \in \mathcal{K}$ such that the measures $\mu_y : \mu_y(B) = (F(B)y, y)$, $B \in \mathcal{B}$, are all absolutely continuous with respect to μ. If E and F are unitarily equivalent then any isometry S from $\mathcal{H}$ to $\mathcal{K}$ which establishes a unitary equivalence between E and F also establishes a unitary equivalence between the restriction of E to $\mathcal{H}_\mu$ and the restriction of F to $\mathcal{K}_\mu$.

1.32 (continued). Select $x_1, x_2, x_3, \ldots$, a complete orthonormal set in $\mathcal{H}$. Put $z_1 = x_1$. Let $\mathcal{H}_1 = \mathcal{H}_{z_1} = \mathcal{H}_{x_1}$, the subspace spanned by $\{E(A)z_1 : A \in \mathcal{B}\}$. Write $P_1 = P_{x_1} = P_{z_1} = $ orthogonal projection on $\mathcal{H}_{z_1}$. Let $z_2 = x_2 - P_{z_1} x_1$ and let $\mathcal{H}_{z_2}$ and P_{z_2} be the subspace and projection corresponding to z_2 defined similarly. Having defined $z_1, z_2, \ldots, z_r, \mathcal{H}_{z_1}, \mathcal{H}_{z_1}, \cdot, \mathcal{H}_{z_r}$ and $P_{z_1}, P_{z_2}, \ldots, P_{z_r}$, write

$$
\begin{aligned}
z_{r+1} &= x_{r+1} - P_{z_1}x_{r+1} - \cdots - P_{z_r}x_{r+1}, \\
\mathcal{H}_{z_{r+1}} &= \text{subspace spanned by } \{E(A)z_{r+1} : A \in \mathcal{B}\}, \\
P_{z_{r+1}} &= \text{Projection on } \mathcal{H}_{z_{r+1}}.
\end{aligned}
$$

Note that each $\mathcal{H}_{z_i}$ is invariant under all $E(B), B \in \mathcal{B}$ and $\mathcal{H}_{z_i} \perp \mathcal{H}_{z_l}$ whenever $i \neq l$. We have

$$
\mathcal{H} = \mathcal{H}_{z_1} \oplus \mathcal{H}_{z_2} \oplus \cdots \oplus \mathcal{H}_{z_n} \oplus \cdots,
$$

since for each $r, x_r \in \mathcal{H}_{z_1} \oplus \cdots \oplus \mathcal{H}_{z_r}$. For each i the restriction of E to $\mathcal{H}_{z_i}$ is of multiplicity one; z_i being its cyclic vector. Further if $\mu_{z_i}(B) = (E(B)z_i, z_i)$, then there exists an invertible isometry S_i from $\mathcal{H}_{z_i}$ onto $L^2(X, \mathcal{B}, \mu_{z_i})$ such that $S_i E(A) S_i^{-1} f = 1_A f$; one such isometry is given by sending $E(A)z_i$ to 1_A and extending by linearity. Define $S : \mathcal{H} \leftrightarrow \sum_{i=1}^{\infty} L^2(X, \mathcal{B}, \mu_{z_i})$ by requiring that S, when restricted to $\mathcal{H}_{z_i}$ is S_i. Then $SE(A)S^{-1}f = 1_A f, f \in \sum_{i=1}^{\infty} L^2(X, \mathcal{B}, \mu_{z_i})$. This isometry suffers from the defect that $\mu_{z_1} \gg \mu_{z_2} \gg \mu_{z_3} \gg \cdots$ need not hold as claimed in the Hahn-Hellinger theorem. We will remove this defect in **1.33**. First let us write

$$
z = \frac{1}{2}z_1 + \frac{1}{2^2}z_2 + \frac{1}{2^3}z_3 + \cdots. \quad \text{Then} \quad \mu_z = \frac{1}{2}\mu_{z_1} + \frac{1}{2^2}\mu_{z_2} + \frac{1}{2^3}\mu_{z_3}\cdots.
$$

Further $E(A) = 0$ if and only if for all i, $\mu_{z_i}(A) = 0$, if and only if $\mu_z(A) = 0$. The measure class of μ_z is therefore the maximal spectral type of E. Thus we have shown that given any spectral measure E defined on a complex separable Hilbert space $\mathcal{H}$, there exists $z \in \mathcal{H}$ such that μ_z is a maximal spectral type of E. We also note, in view of exercise **1.14**, that given a finite measure ν on $(X, \mathcal{B})$ absolutely continuous with respect to the maximal spectral type of E, there exists $x \in \mathcal{H}$ such that $\mu_x = \nu$.

The Hahn-Hellinger Theorem (First Form)

1.33. Given $x \in \mathcal{H}$, let ν be a finite measure singular to μ_x, and such that $\mu_x + \nu$ is a maximal spectral type of E. If $y \in \mathcal{H}$ is such that $\mu_y = \nu$ then $y \perp \mathcal{H}_x$. Further if $z = x + y$, then $\mathcal{H}_z = \mathcal{H}_x \oplus \mathcal{H}_y$ so that $x \in \mathcal{H}_z$. Thus given $x \in \mathcal{H}$, there exists $z \in \mathcal{H}$ such that $x \in \mathcal{H}_z$ and μ_z is the maximal spectral type of E. We are now in a position to prove:

1.34 Hahn-Hellinger Theorem (First Form). *Let E be a spectral measure on $(X, \mathcal{B})$ acting in a complex separable Hilbert space $\mathcal{H}$. Then*

(i) *there exist σ-finite measures $\mu_1 \gg \mu_2 \gg \mu_3 \gg \cdots$, and an invertible isometry:*

$$
S : \mathcal{H} \leftrightarrow \sum_{i=1}^{\infty} L^2(X, \mathcal{B}, \mu_i)
$$

such that $SE(A)S^{-1}f = 1_A f, A \in \mathcal{B}, \quad f \in \sum_{i=1}^{\infty} L^2(X, \mathcal{B}, \mu_i).$

(ii) *If $\nu_1 \gg \nu_2 \gg \nu_3 \cdots$ is another such sequence of measures, i.e., for which there exists an invertible isometry*

$$S_1 : \mathcal{H} \leftrightarrow \sum_{i=1}^{\infty} L^2(X, \mathcal{B}_i, \nu_i)$$

with

$$S_1 E(A) S_1^{-1} f = 1_A f, A \in \mathcal{B}, \quad f \in \sum_{i=1}^{\infty} L^2(X, \mathcal{B}, \nu_i),$$

then for each i, μ_i and ν_i are in the same measure class.

Proof. Let $x_1, x_2, x_3, \ldots$, be a complete orthonormal set in $\mathcal{H}$, or any countable set whose closed linear span is $\mathcal{H}$. By **1.33** choose z_1 such that μ_{z_1} is a maximal spectral type of E and $x_1 \in \mathcal{H}_{z_1}$. Put $\mu_{z_1} = \mu_1$, $\mathcal{H}_{z_1} = \mathcal{H}_1$. Let $x_2' = x_2 - P_{\mathcal{H}_1} x_2$, where $P_{\mathcal{H}_1}$ denotes the orthogonal projection on $\mathcal{H}_1$. Choose $z_2 \in \mathcal{H}_1^{\perp}$, such that $\mu_2 = \mu_{z_2}$ is a maximal spectral type of the restriction of E to $\mathcal{H}_1^{\perp}$, and $x_2' \in \mathcal{H}_{z_2} = \mathcal{H}_2$. Proceeding thus we get $z_1, z_2, z_3, \ldots$, such that

(i) $\mathcal{H}_{z_i} \perp \mathcal{H}_{z_j}$ if $i \neq j$,
(ii) $x_n \in \mathcal{H}_{z_1} \oplus \cdots \oplus \mathcal{H}_{z_n}$,
(iii) $\mathcal{H} = \sum_{i=1}^{\infty} \mathcal{H}_{z_i}$ (this follows from (ii)),
(iv) $\mu_{z_1} \gg \mu_{z_2} \gg \mu_{z_3} \gg \cdots$.

Now there exist isometries $S_j : \mathcal{H}_j \leftrightarrow L^2(X, \mathcal{B}, \mu_j)$ $(\mu_j = \mu_{z_j})$, such that $S_j E(B) S_j^{-1} f = 1_B f, B \in \mathcal{B}, f \in L^2(X, \mathcal{B}, \mu_j)$.

We define $S : \mathcal{H} \leftrightarrow \sum_{j=1}^{\infty} L^2(X, \mathcal{B}, \mu_j)$ by requiring that the restriction of S to $\mathcal{H}_j$ be S_j. Then

$$SE(B)S^{-1} f = 1_B f, \quad \text{for all} \quad B \in \mathcal{B} \text{ and } f \in \sum_{j=1}^{\infty} L^2(X, \mathcal{B}, \mu_j).$$

Thus (i) is proved.

Now suppose that $\nu_1 \gg \nu_2 \gg \cdots$, is another sequence of measures such that there exists an isometry S_1 from $\mathcal{H}$ onto $\sum_{j=1}^{\infty} L^2(X, \mathcal{B}, \nu_j)$ satisfying

$$S_1 E(B) S_1^{-1} f = 1_B f \text{ for all } B \in \mathcal{B}, f \in \sum_{j=1}^{\infty} L^2(X, \mathcal{B}, \nu_j).$$

We have to show that for each i, μ_i and ν_i have the same null sets. There is no loss of generality if we assume that there exist sets $X = A_1 \supseteq A_2 \supseteq A_3 \cdots$, and that $\mu_i = \mu_1 \mid_{A_i}$. Similarly we may assume that there exist sets $X = B_1 \supseteq B_2 \supseteq B_3 \supseteq \cdots$, and that $\nu_i = \nu_1 \mid_{B_i}$. Now $\mu_1(A) = 0$ if and only if $E(A) = 0$ if and only if $\nu_1(A) = 0$, so that μ_1 and ν_1 have the same null sets. Suppose we have proved that μ_i and ν_i have the same null sets for $1 \leq i \leq r$. Consider μ_{r+1} and ν_{r+1}. Assume, in order to arrive at a contradiction, that there is a set A such that $\mu_{r+1}(A) > 0$ and $\nu_{r+1}(A) = 0$. Let $k > r + 1$ be the first integer, if there is one, such that $\mu_k \mid_A$ is not in the measure class of $\mu_{k-1} \mid_A$. In such a case write B for a subset of A with $\mu_{k-1}(B) > 0$ and $\mu_k(B) = 0$. Put $B = A$ if there is no such k. Let λ be the measure class of $\mu_1 \mid_B (= \nu_1 \mid_B)$ and let $\mathcal{H}_\lambda = \{z \in \mathcal{H} : \lambda \gg \mu_z\}$. Then $\mathcal{H}_\lambda$ is a subspace of $\mathcal{H}$ invariant under $E(B), B \in \mathcal{B}$. Now

$$SH_\lambda = \left\{ f \in \sum_{j=1}^{\infty} L^2(X, \mathcal{B}, \mu_j) : f \text{ vanishes outside } B \right\}$$

$$= \sum_{j=1}^{\infty} L^2(X, \mathcal{B}, \mu_j \mid_B) \underset{\text{def}}{=} \mathcal{K},$$

where the sum has $(k - 1)$ copies of $L^2(X, \mathcal{B}, \mu_1 \mid_B)$ if $k < \aleph_0$, otherwise it has $\aleph_0$ copies. The isometry S_1 maps $\mathcal{H}_\lambda$ onto $rL^2(X, \mathcal{B}, \nu_1 \mid_B) = \mathcal{L}$ say. The spectral measures on $\mathcal{K}$ and $\mathcal{L}$ defined by

$$F_\mu(C)f = 1_C f, \quad f \in \mathcal{K}, \quad C \in \mathcal{B},$$

$$F_\nu(C)g = 1_C g, \quad g \in \mathcal{L}, \quad C \in \mathcal{B},$$

are then unitarily equivalent, each being equivalent to the restriction of E to $\mathcal{H}_\lambda$. It follows from **1.31** that the number of copies of $L^2(X, \mathcal{B}, \mu_1 \mid_B)$ in $\mathcal{K}$ must be r, which is the number of copies of $L^2(X, \mathcal{B}, \nu_1 \mid_B)$ in $\mathcal{L}$. This is a contradiction since the number of copies of $L^2(X, \mathcal{B}, \mu_1 \mid_B)$ in $\mathcal{K}$ is bigger than or equal to $r+1 > r$. Hence μ_{r+1} and ν_{r+1} have the same null sets. This proves the theorem completely.

The Hahn-Hellinger Theorem (Second Form)

1.35. There is another form in which the Hahn-Hellinger theorem can be stated. This form is more useful in the applications we have in mind. Also it is this form of the theorem which generalises to the case of non-separable Hilbert spaces.

1.36 Hahn-Hellinger Theorem (Second Form). *Let E be a spectral measure on $(X, \mathcal{B})$ acting in a complex separable Hilbert space $\mathcal{H}$. Then there exist mutually singular σ-finite measures $\nu_\infty, \nu_1, \nu_2, \ldots$ and an invertible isometry*

$$S : \mathcal{H} \leftrightarrow L^2(X, \mathcal{B}, \nu_\infty, l_2) \oplus \sum_{n=1}^{\infty} nL^2(X, \mathcal{B}, \nu_n),$$

such that for all $A \in \mathcal{B}$, $f \in L^2(X, \mathcal{B}, \nu_\infty, l_2) \oplus \sum_{n=1}^{\infty} nL^2(X, \mathcal{B}, \nu_n)$,

$$SE(A)S^{-1}f = 1_A f.$$

If $\nu_1', \nu_2', \nu_3', \ldots$, is another such sequence of mutually singular measures then for each i, ν_i and ν_i' have the same null sets.

Proof. By the first form of the Hahn-Hellinger theorem there exist a finite measure μ, sets $X = A_1 \supseteq A_2 \supseteq A_3 \cdots$, in $\mathcal{B}$, and an isometry S from $\mathcal{H}$ onto $\sum_{n=1}^{\infty} L^2(X, \mathcal{B}, \mu \mid_{A_i})$ satisfying $SE(A)S^{-1}f = 1_A f$. Let $A_{\infty} = \cap_{i=1}^{\infty} A_i$. We have

$$L^2(X, \mathcal{B}, \mu \mid_{A_n}) = \sum_{k=n}^{\infty} (L^2(X, \mathcal{B}, \mu \mid_{(A_k - A_{k+1})})) \oplus L^2(X, \mathcal{B}, \mu \mid_{A_{\infty}}).$$

Therefore

$$\sum_{n=1}^{\infty} L^2(X, \mathcal{B}, \mu \mid_{A_n}) = \sum_{n=1}^{\infty} \sum_{k=n}^{\infty} (L^2(X, \mathcal{B}, \mu \mid_{(A_k - A_{k+1})})) \oplus L^2(X, \mathcal{B}, \mu \mid_{A_{\infty}}).$$

Now $L^2(X, \mathcal{B}, \mu \mid_{(A_k - A_{k+1})})$ occurs k times in the above summation and $L^2(X, \mathcal{B}, \mu \mid_{A_{\infty}})$ occurs $\aleph_0$ times hence we have

$$\sum_{n=1}^{\infty} L^2(X, \mathcal{B}, \mu_n) = L^2(X, \mathcal{B}, \nu_{\infty}, l_2) \oplus \sum_{k=1}^{\infty} k L^2(X, \mathcal{B}, \nu_k),$$

where $\nu_{\infty} = \mu \mid_{A_{\infty}}$ and, ν_k is $\mu \mid_{(A_k - A_{k+1})}$. Clearly $SE(A)S^{-1}f = 1_A f$ for $f \in L^2(X, \mathcal{B}, \nu_{\infty}, l_2) \oplus \sum_{n=1}^{\infty} n L^2(X, \mathcal{B}, \nu_n)$.

If $\nu_1', \nu_2', \nu_3', \ldots$ is another such sequence of mutually singular measures then the measure $\sum_{k=n}^{\infty} \nu_k'$ has the same null sets as $\mu_n = \mu \mid_{A_n}, n = 1, 2, 3, \ldots$. From this it follows that ν_k and ν_k' have the same null sets, being in the measure class of $\mu \mid_{(A_k - A_{k+1})}$, $k = 1, 2, 3, \ldots$. This completes the proof.

1.37. For any measure ν, let $[\nu]$ denote the measure class of ν. Let $[\nu_{\infty}], [\nu_1], \ldots,$ be the sequence of mutually singular measure classes associated to E as per the second form of Hahn-Hellinger theorem. We say that E has multiplicity $\aleph_0$ or E has infinite multiplicity if $[\nu_{\infty}] \neq [0]$. We say that E has multiplicity $n < \aleph_0$ if $[\nu_n] \neq [0]$ and $[\nu_k] = [0]$ for all $k \geq n + 1$. We say that E has uniform multiplicity $n \leq \aleph_0$ if $[\nu_n] \neq [0]$ and $[\nu_k] = [0]$ for all $k \neq n$. Note that the sequence of measure classes $[\nu_{\infty}], [\nu_1], [\nu_2] \ldots,$ associated with E as per the second form of the Hahn-Hellinger theorem, is a complete invariant of unitary equivalence.

1.38. If X is a group and $\mathcal{B}$ is invariant under $x \to x^{-1}$, $x \to ax$, for all $a \in X$ then we can define new spectral measures $\tilde{E} : A \to E(A^{-1})$, $E_a : A \to E(aA)$. Clearly the measure classes associated with $\tilde{E}$ as per the second form of the Hahn-Hellinger theorem are $[\tilde{\nu}_k]$, $1 \leq k \leq \infty$, where $\tilde{\nu}_k(A) = \nu_k(A^{-1}), A \in \mathcal{B}$. Moreover E and $\tilde{E}$ are unitarily equivalent if and only if for each k, $[\nu_k]$ and $[\tilde{\nu}_k]$ are the same measure class. Similarly E and E_a are unitarily equivalent if and only if for each k, the measure classes $[\nu_k]$ and $[\nu_{k,a}]$ are the same, $\nu_{k,a}$ being the translate of ν by a. We will say that E is symmetric if E and $\tilde{E}$ are unitarily equivalent.

Exercise. Let f be a Borel measurable real valued function on the unit interval I such that for every real x, $f^{-1}(x)$ is a finite set. Let l denote the Lebesgue measure on I. Show that I can be decomposed into pairwise disjoint Borel sets $A_0, A_1, A_2, \ldots$, such that $l(A_0) = 0$ and f is k to one on A_k for each $k \geq 1$. Some of the A_k's may be empty. The images under f of $A_k, k = 1, 2, \ldots$, are pairwise disjoint. Show further that each A_k can be decomposed into pairwise disjoint Borel sets $A_{k,i}, 1 \leq i \leq k$ on each of which f is one-one and the restrictions of f to $A_{k,i}, 1 \leq i \leq k$ induce measures on $f(A_{k,i}), 1 \leq i \leq k$, which are mutually absolutely continuous. Assume now that f is bounded and consider the Hermitian operator H on $L^2[0,1]$: $Hg = f \cdot g$, $g \in L^2[0,1]$. Show that the above considerations describe completely the spectral measure of H.

Representation of Second Order Stochastic Processes

1.39. We now complete the discussion on the representation of second order stochastic processes by treating the case of arbitrary multiplicity. Let $(X_t)_{t \in \mathbb{R}}$ be a purely non-deterministic second order stochastic process, $E(X_t) = 0$, for all t. Let $\mathcal{M}_\infty =$ closure of $\bigcup_{t \in \mathbb{R}} \mathcal{M}_t$ and let P_t denote the orthogonal projection from $\mathcal{M}_\infty$ on the closed subspace $\mathcal{M}_t$ spanned by $\{X_s : s \leq t\}$. Let $Q_t = \lim_{s \to t, s > t} P_s$. Let E be the spectral measure given by $E(a,b] = Q_b - Q_a$. Let $z_1, z_2, \ldots$, be vectors in $\mathcal{M}_\infty$ such that $\mathcal{H}_{z_1}, \mathcal{H}_{z_2}, \ldots$, are mutually orthogonal and their direct sum is $\mathcal{M}_\infty$. Further we may assume, if necessary, that $\mu_{z_1} \gg \mu_{z_2} \gg \mu_{z_3}, \cdots$. We can write

$$X_t = X_{t,1} + X_{t,2} + X_{t,3} + \cdots + X_{t,n} + \cdots,$$

where $X_{t,i}$ is the orthogonal projection of X_t on $\mathcal{H}_{z_i}$. Put $z_i(t) = Q_t z_i$. Then $X_{t,i}$ may be written in the form

$$X_{t,i} = \int_{-\infty}^{\infty} f_{t,i}(s) z_i(ds).$$

Moreover, since $X_t \in Q_t \mathcal{M}_\infty$, $f_{i,t}(s) = 0$ for $s > t$. Thus

$$X_{t,i} = \int_{-\infty}^{t} f_{t,i}(s) z_i(ds).$$

We have thus proved that every purely non-deterministic second order stochastic process $(X_t)_{t \in \mathbb{R}}, E(X_t) = 0$ for all t, has a representation of the form

$$X_t = \sum_{i=1}^{\infty} \int_{-\infty}^{t} f_{t,i}(s) z_i(ds),$$

where for each i, $z_i(t), t \in \mathbb{R}$, is a process with orthogonal increments such that:

(i) $(z_i(b) - z_i(a)) \perp (z_j(d) - z_j(c))$ whenever $j \neq i$,
(ii) if μ is the measure defined by $\mu_i(a,b] = \| z_i(b) - z_i(a) \|^2$ then $\mu_i \gg \mu_{i+1}$.

In connection with the contents of this section and for more details see H. Cramér [1] and T. Hida [4].

1.40. For other accounts of Hahn-Hellinger theorem we refer to M. H. Stone [6], H. Helson [3]. For spectral multiplicity theory on non-separable Hilbert spaces see P. R. Halmos [2], A. I. Plessner and V. A. Rokhlin [5].

Exercise. Let I denote the unit interval with Lebesgue measure. Define f on $I \times I$ by $f(x, y) = x$. Show that the spectral measure of the Hermitian operator $H : g \to f \cdot g$, $g \in L^2(I \times I)$, has uniform multiplicity $\aleph_0$ with maximal spectral type Lebesgue.

Chapter 2

The Spectral Theorem for Unitary Operators

The Spectral Theorem: Multiplicity One Case

2.1. In this chapter we briefly discuss the spectral theorem for unitary operators.

2.2. Let U be a unitary operator on a Hilbert space $\mathcal{H}$. Let $x \in \mathcal{H}$, then the function $r(n) = (U^n x, x), n \in \mathbb{Z}$, is positive definite in the sense that for any finite set $c_1, c_2, \ldots, c_n$ of complex numbers

$$\sum_{i=1}^{n} \sum_{j=1}^{n} c_i r(i-j) \overline{c_j} = \left\| \sum_{i=1}^{\infty} c_i U^i x \right\|^2 \geq 0.$$

A theorem of Herglotz states that any positive definite function on the integers is the Fourier transform of a finite non-negative measure on the circle group S^1. So we can write the function r in the form

$$r(n) = \int_{S^1} z^{-n} d\mu, \quad n \in \mathbb{Z},$$

where μ is a finite measure on the unit circle S^1.

Let $\mathcal{H}_x$ denote the closed subspace of $\mathcal{H}$ spanned by $\{U^n x : n \in \mathbb{Z}\}$. The linear manifold $\mathcal{M}$ of finite linear combinations of $U^k x, k \in \mathbb{Z}$ is dense in $\mathcal{H}_x$. With the finite linear combination $\sum_{k=-n}^{n} c_k U^k x$ we can associate the trigonometric polynomial $\sum_{k=-n}^{n} c_k z^{-k} \in L^2(S^1, \mu)$. Then

$$\left(\sum_{k=-n}^{n} c_k U^k x, \sum_{l=-m}^{m} d_l U^l x \right) = \int_{S^1} \sum_{k=-n}^{n} c_k z^{-k} \sum_{l=-m}^{m} \overline{d_l} z^l d\mu,$$

= the inner product of the trigonometric polynomials

$$\sum_{k=-n}^{n} c_k z^{-k} \quad \text{and} \quad \sum_{l=-m}^{m} d_l z^{-l} \quad \text{in} \quad L^2(S^1, \mu).$$

Let ψ denote the map which sends each finite linear combination $\sum_{k=-n}^{n} c_k U^k x$ to the associated trigonometric polynomial $\sum_{k=-n}^{n} c_k z^{-k}$. The map ψ, defined on $\mathcal{M}$, is invertible, linear, and inner product preserving. Its image is the linear manifold of trigonometric polynomials in $L^2(S^1, \mu)$ which is dense in $L^2(S^1, \mu)$. The map ψ therefore extends to an invertible isometry from $\mathcal{H}$ onto $L^2(S^1, \mu)$. We continue to denote the extended isometry by ψ.

Assume for the time being that $\mathcal{H}_x = \mathcal{H}$. If F denotes the spectral measure on $L^2(S^1, \mu)$ defined by $F(A) =$ multiplication by 1_A, then F has multiplicity one, and hence the spectral measure $E = \psi^{-1} F \psi$ has multiplicity one. If $y_1, y_2 \in \mathcal{H}$ and if $f_1 = \psi(y_1), f_2 = \psi(y_2)$ are the corresponding elements in $L^2(S^1, \mu)$ then

$$(U^n y_1, y_2) = \int_{S^1} z^{-n} f_1 \overline{f}_2 d\mu = \int_{S^1} z^{-n} d\nu,$$

where ν is the complex-valued measure $f_1 \overline{f}_2 d\mu$ which is same as the complex-valued measure $\nu(\cdot) = (F(\cdot)f_1, f_2) = (E(\cdot)y_1, y_2)$. We express $(U^n y_1, y_2)$ in the form

$$(U^n y_1, y_2) = \int_{S^1} z^{-n} (E(dz)y_1, y_2) \tag{1}$$

and write

$$U^n = \int_{S^1} z^{-n} dE, \quad \text{or} \quad U^n = \int_{S^1} z^{-n} E(dz), \tag{2}$$

depending on which is more convenient in a given context. The interpretation of formula (2) is that (1) holds for all $y_1, y_2 \in \mathcal{H}$.

The Spectral Theorem: Higher Multiplicity Case

2.3. In case there is no single vector $x \in \mathcal{H}$ with $\mathcal{H}_x = \mathcal{H}$, we can write $\mathcal{H}$ as an orthogonal direct sum $\sum \mathcal{H}_i$ which may be finite or infinite, such that each $\mathcal{H}_i$ is invariant under U and U^{-1}, and is moreover a cyclic subspace in the sense that it is generated by $(U^k x_i), k \in \mathbb{Z}$, for some $x_i \in \mathcal{H}$. The restriction of U to $\mathcal{H}_i$ admits a "spectral resolution"

$$U^n = \int_{S^1} z^{-n} dE_i$$

in the sense of (1) and (2) above. The spectral measure

$$E = E_1 + E_2 + E_3 + \cdots$$

satisfies

$$U^n = \int_{S^1} z^{-n} dE$$

in the sense that for all $y_1, y_2 \in \mathcal{H}$,

$$(U^n y_1, y_2) = \int_{S^1} z^{-n}(E(dz)y_1, y_2).$$

This is the spectral theorem for U.

2.4. We say that U has simple spectrum if E has multiplicity one, equivalently, there exists $x \in \mathcal{H}$ such that $U^n x, n \in \mathbb{Z}$, span $\mathcal{H}$. By the spectral type or maximal spectral type of U we mean those of E. Similarly, the multiplicity of U means that of E etc. One can formulate these definitions entirely in terms of U. Thus, to say that U has uniform multiplicity n with maximal spectral type μ (a finite measure on S^1) means there exist n vectors $x_1, x_2, \ldots, x_n \in \mathcal{H}$ such that

(i) $U^k x_i \perp U^m x_j$ if $i \neq j$, for all $m, k \in \mathbb{Z}$.
(ii) $(U^k x_i, x_i) = \int_{S^1} z^{-k} d\mu$ for all i and k.
(iii) $\{U^k x_i : -\infty < k < \infty, 1 \leq i \leq n\}$ span $\mathcal{H}$.

In case μ is in the class of Lebesgue measure on S^1, we say that U has uniform Lebesgue spectrum with multiplicity n. One may similarly define uniform multiplicity $\aleph_0$ for U.

2.5. We state the spectral theorem for a group of unitary operators indexed by a locally compact abelian group.

Let G be a locally compact abelian group and $\hat{G}$ its dual. Let

$$U_g, g \in G,$$

be a continuous unitary representation of G. Then there exists a spectral measure E on the Borel subsets of $\hat{G}$ such that for all $g \in G$, $U_g = \int_{\hat{G}} \chi_{-g} dE$ where χ_{-g} represents the character on G corresponding to $-g \in G$. A proof of this can be given along the same lines as above.

Chapter 3

Symmetry and Denseness of the Spectrum

3.1. In this chapter we prove two properties of the spectrum of the unitary operator associated with an aperiodic non-singular automorphism, viz., that it is symmetric and fills the entire unit circle. We also discuss the spectrum of such an operator multiplied by a unitary function.

Spectrum of U_T: It is Symmetric

3.2. Let $(X, \mathcal{B}, m)$ be a standard probability space. Let $T : X \to X$ be a Borel automorphism on $(X, \mathcal{B})$ such that m and the measure $m_T : m_T(A) = m(TA), A \in \mathcal{B}$ have the same null sets. We then say that T is non-singular with respect to m or that m is quasi-invariant under T.

3.3. Let φ be a complex valued measurable function on X of absolute value one. On $L^2(X, \mathcal{B}, m)$ define two unitary operators U_T and V_φ as follows:

$$U_T f(x) = \sqrt{\frac{dm_T}{dm}(x)} f(Tx),$$

$$V_\varphi f(x) = \varphi(x) U_T f(x) = \varphi(x) \sqrt{\frac{dm_T}{dm}(x)} f(Tx),$$

where $f \in L^2(X, \mathcal{B}, m)$. Note that U_T agrees with V_φ if $\varphi(x) = 1$ for all x (mod m). Let

$$U_T^n = \int_{S^1} z^{-n} dE.$$

Now U_T takes real functions into real functions, indeed U_T takes positive functions into positive functions, hence

$$\overline{(U^n f, f)} = (U^n \overline{f}, \overline{f}) = \int_{S^1} z^{-n} (E(dz) \overline{f}, \overline{f}), \quad n \in \mathbb{Z},$$

$$\overline{(U^n f, f)} = \int_{S^1} z^n (E(dz) f, f) = \int_{S^1} z^{-n} (\tilde{E}(dz) f, f), \quad n \in \mathbb{Z},$$

where $\tilde{E}(A) = E(A^{-1})$, $A \subseteq S^1$. It is easy to see from this that the measure classes associated to E and $\tilde{E}$ as per the second form of the Hahn-Hellinger theorem are the same, so that E and $\tilde{E}$ are unitarily equivalent, which means, by definition, that E is symmetric. (See **1.38**).

The spectral measure of V_φ need not be symmetric as will be shown later.

Spectrum of U_T: It is Dense

3.4. We will now show that if T is aperiodic, then the spectrum of V_φ is the entire unit circle, where by the spectrum of V_φ we mean the collection of $\lambda \in \mathbb{C}$ such that $V_\varphi - \lambda I$ is not invertible. In particular it will follow that the spectrum of U_T, when T is aperiodic, is the entire unit circle.

3.5. Let the hypothesis be as in **3.4**. To show that spectrum of V_φ is all of S^1 we must show that each $\lambda \in S^1$ is an approximate eigenvalue of V_φ, equivalently, we must show that given $\lambda \in S^1$ and $\varepsilon > 0$ there exists an $f \in L^2(X, \mathcal{B}, m)$ of norm one such that

$$\| V_\varphi f - \lambda f \| < \varepsilon.$$

3.6. Fix $\varepsilon > 0$ and choose δ positive but $< \varepsilon^2/4$. Let n be a positive integer with $1/n < \varepsilon^2/4$. Since T is aperiodic, by Rokhlin's lemma we can choose, for the given $\delta > 0$ and n, a measurable set A such that $A, TA, \ldots, T^{n-1}A$ are pairwise disjoint and

$$m\left(X - \bigcup_{k=0}^{n-1} T^k A\right) < \delta.$$

Write $C = X - \bigcup_{k=0}^{n-1} T^k A$. Our definition of f is dictated by the requirement that $V_\varphi f$ should equal λf on a large set. We set f equal to a constant a on A. We will choose a presently. Inductively define:

$$f(x) = \lambda \overline{\varphi}(T^{-1} x) \sqrt{\left(\frac{dm_{T^{-1}}}{dm}(x)\right)} f(T^{-1} x), \quad for \ x \in T^k A, 1 \le k \le n-1,$$

and finally set f equals 1 on C. ($\overline{\varphi}$ denotes the complex conjugate of φ). It is easy to see that

$$\int_{T^k A} |f|^2 \, dm = \int_{T^{k-1} A} |f|^2 \, dm.$$

Hence $|| f ||^2 = na^2 m(A) + m(C)$. Now choose

$$a = + \left(\frac{1 - m(C)}{n \cdot m(A)} \right)^{1/2} \quad \text{so} \quad \text{that} \quad || f ||^2 = 1.$$

Note that

$$(V_\varphi f)(x) = \lambda \cdot f(x) \quad \text{for} \quad x \in \bigcup_{k=0}^{n-2} T^k A \tag{1}$$

and

$$\int_{T^k A} | f |^2 \, dm \le 1/n, \quad 0 \le k \le n - 1.$$

In particular

$$\int_A | f |^2 \, dm \le 1/n, \quad \int_{T^{n-1} A} | f |^2 \, dm \le 1/n. \tag{2}$$

Finally using (1) and (2) we get

$$|| V_\varphi f - \lambda f ||^2 = \int_{(T^{n-1} A) \cup C} | V_\varphi f - \lambda f |^2 \, dm \le 2 \left(\frac{1}{n} + m(C) \right) < \varepsilon^2.$$

Thus every $\lambda \in S^1$ is an approximate eigenvalue of V_φ.

3.7. Assume that T is measure preserving and consider $U = U_T$. We say that a function f on X is an ε-eigenfunction with eigenvalue λ if $|| f || = 1$ and $|| Uf - \lambda f || < \varepsilon$. It is clear from our construction above that for a given ε and λ the function f_λ which is 1 on A, λ on TA,..., λ^{n-1} on $T^{n-1} A$ and 1 on C will be an ε-eigenfunction of U_T with eigenvalue λ. (Here A and C are as in **3.4** with $\delta < \varepsilon^2/4$, $\frac{1}{n} < \varepsilon^2/4$). Moreover $| f_\lambda | = 1$ and $f_\lambda f_\mu = f_{\lambda \mu}$. This last property is worth noting since the product of two ε-eigenfunctions in the case of a general unitary operator need not be an ε-eigenfunction.

3.8. If the function φ is of the form $\frac{\xi \circ T}{\xi}$ for some measurable ξ of absolute value 1, then V_φ and U are unitarily equivalent. For if we set $Wf = \xi f$, $f \in L^2(X, \mathcal{B}, m)$, then W is unitary and we see that

$$V_\varphi f = \varphi U_T f = \frac{\xi \circ T}{\xi} U_T f = \xi^{-1} U_T \xi f = W^{-1} U_T W f,$$

whence V_φ and U_T are unitarily equivalent.

3.9. If T is ergodic and measure preserving and U_T and V_φ are unitarily equivalent then φ is a coboundary, i.e., ϕ is of the form $\frac{\xi \circ T}{\xi}$ for some measurable function ξ of absolute value one. For when T is measure preserving, since $m(X) < \infty$, U_T admits 1 as an eigenvalue. If U and V_φ are unitarily equivalent then 1 is an eigenvalue of V_φ also. Hence there is a function f such that $V_\varphi f = f$ a.e., i.e., $\varphi \cdot f \circ T = f$ a.e. Since $\mid \varphi \mid = 1$, we see that $\mid f \circ T \mid = \mid f \mid$. By ergodicity of T, $\mid f \mid$ is a constant which we may assume to be equal to 1. If we set $\xi = \frac{1}{f}$ we see that $\varphi = \frac{\xi \circ T}{\xi}$.

3.10. Assume that T is ergodic and non-singular. It does not seem to be known whether unitary equivalence of U_T and V_φ implies that φ is a coboundary.

Exercise. Show that if T is ergodic and V_φ admits an eigenvalue, then φ is a coboundary and there is a measure which is finite, invariant under T, and has the same null sets as m.

Examples

3.11 Example. Let $X = S^1$ and $Tx = \alpha x$, where $\alpha = e^{2\pi i a}$, a being irrational. Let m be Lebesgue measure on S^1. Then T is measure preserving and ergodic. For each $n \in \mathbb{Z}$ the function $f_n(x) = x^n$ is an eigenvector of U_T with eigenvalue α^n. These eigenvectors form a complete orthonormal set for $L^2(S^1, \mathcal{B}, m)$. Let $\beta \in S^1$, $\beta \neq \alpha^n$ for any $n \in \mathbb{Z}$, and set $\varphi(x) = \beta$ for all $x \in S^1$. Then

$$V_\varphi f_n = \beta f_n \circ T = \beta \alpha^n f_n.$$

Thus f_n is an eigenfunction of V_φ with eigenvalue $\beta \alpha^n$, and the eigenfunctions $f_n, n \in \mathbb{Z}$, form a complete orthonormal set. Since β is not equal to α^k for any k, we conclude that V_φ and U_T are not unitarily equivalent. If β is equal to α^n for some n, then V_φ and U_T are unitarily equivalent since they have the same set of eigenvalues, each with multiplicity one, and their eigenvectors form a complete orthonormal set. The function $\varphi(x) = \beta$ is a coboundary (with respect to $T : x \to \alpha x$) if and only if β is in the group generated by α. (See H. Helson [3].) Note that if $\beta \notin \{\alpha^n : n \in \mathbb{Z}\}$, then the spectral measure of V_φ is not symmetric since it is supported on $\{\beta \alpha^n\}$, $n \in \mathbb{Z}$.

3.12 Example. Let T be as in the above example. Let

$$\varphi(z) = z^p, z \in S^1, \; p \text{ a positive integer.}$$

Let V stand for V_φ. Then

$$(V^n f)(z) = \begin{cases} z^p(\alpha z)^p \cdots (\alpha^{n-1} z)^p f(\alpha^n z) & \text{if } n > 0, \\ f(z) & \text{if } n = 0, \\ (\alpha^{-1} z)^{-p} \cdots (\alpha^{-n} z)^{-p} f(\alpha^n z), & \text{if } n < 0, \end{cases}$$

or

$$(V^n f)(z) = \begin{cases} \alpha^{\frac{1}{2} pn(n-1)} z^{np} f(\alpha^n z), & \text{if } n > 0, \\ f(z) & \text{if } n = 0, \\ \alpha^{-\frac{1}{2} pn(n+1)} z^{np} f(\alpha^n z), & \text{if } n < 0. \end{cases}$$

If we set $f_1 = 1$, $f_2 = z, \ldots, f_p = z^{p-1}$, then

(i) $(V^n f_i, V^m f_i) = 0$ if $m \neq n$,
(ii) $(V^n f_i, V^m f_j) = 0$ if $i \neq j$ for all m, n,
(iii) $\{V^n f_i : n \in \mathbb{Z}, \quad i = 1, 2, \ldots, p\}$ span $L^2(S, \mathcal{B}, m)$.

This shows that $V_\varphi = V$ has uniform Lebesgue spectrum with multiplicity p.

3.13. Let T be as in the above example. Let φ be an inner function. Then V_φ has uniform Lebesgue spectrum. The multiplicity is finite if φ is a finite Blaschke product, otherwise the multiplicity is $\aleph_0$. We see this as follows: Let $H^2 \subseteq L^2(S^1, \mathcal{B}, m)$ be those functions whose negative Fourier coefficients vanish. Now $U_T H^2 = H^2$ as can be readily verified. Hence $V_\varphi H^2 = \varphi U_T H^2 = \varphi H^2$. If $K = H^2 \ominus \varphi H^2$ then

$$(V_\varphi)^n K \perp (V_\varphi)^m K$$

for $m \neq n$, and the direct sum of $(V_\varphi)^n K$ over all $n \in \mathbb{Z}$ is $L^2(S^1, \mathcal{B}, m)$. This shows that V_φ has uniform Lebesgue spectrum with multiplicity equal to the dimension of K, which equals $n < \aleph_0$ if φ is a finite Blaschke product with n factors, and equal to $\aleph_0$ otherwise. (See S. C. Bagchi, J. Mathew, M. G. Nadkarni [1].)

3.13 (continued). Let $z = e^{ix}$, x real, and let $\phi : S^1 \to S^1$ be a smooth map. Such a ϕ can be represented as

$$\phi(z) = e^{2\pi i \tilde{\phi}(x)} \cdot e^{2\pi i m x}$$

where $\tilde{\phi}(x) : \mathbb{R} \to \mathbb{R}$ is periodic with period one and smooth. In this representation, $m \in \mathbb{Z}$ is unique, while $\tilde{\phi}$ is unique up to an additive integer constant. The number m is called the degree $d(\phi)$ of ϕ. It is known that if $m = 0$ and $\tilde{\phi}$ is absolutely continuous then the maximal spectral type of V_ϕ is singular; in contrast, as soon as $m = d(\phi) \neq 0$ and $\tilde{\phi}'$ is of bounded variation, V_ϕ has Lebesgue spectrum (see A. Iwanik, M. Lemańczyk, D. Rudolph [4]).

Let ψ be a real valued measurable function on the unit interval,

$$\varphi = e^{2\pi i \psi},$$

and let T be as above. In a recent paper Mélanie Guenais [2] has shown that the multiplicity of V_φ is related to the total variation of ψ. The multiplicity of V_φ is majorised by $\max(2, \frac{2\pi}{3} Var(\psi))$. If ψ is absolutely continuous, then the multiplicity of V_φ is majorised by $\max(2, |\int_0^1 \psi'(x)dx| + 1)$. The bound is attained for $\psi(x) = nx$.

3.14. Let T be as in the above example. Define φ by

$$\varphi(e^{2\pi i x}) = -1 \quad \text{if} \quad 0 \leq x < \beta, \quad \varphi(e^{2\pi i x}) = 1 \quad \text{if} \quad \beta \leq x < 1.$$

The spectrum of V_φ has been discussed by G. W. Riley [6]. It is shown that for almost every β with respect to Lebesgue measure V_φ has simple continuous singular spectrum. Further, for suitable α and β, the maximal spectral type μ of V_φ has the property that $\mu * \mu$ and μ are mutually singular. This latter fact was proved earlier by Katok and Stepin [5] where it was used to disprove a conjecture of Kolmogorov which asserted that the maximal spectral type μ of U_T, where T is ergodic and measure preserving, has the property that $\mu * \mu$ is absolutely continuous with respect to μ.

Chapter 4

Multiplicity and Rank

A Theorem on Multiplicity

4.1. The rank of a measure preserving automorphism σ is greater than or equal to the spectral multiplicity of U_σ. This is true for a class of non-singular automorphisms as well. We will suitably adapt the exposition given in Chacon [1], which in turn is an improvement of an earlier work by Katok and Stepin [4].

4.2 Let E be a spectral measure on a Borel space $(X, \mathcal{B})$ acting in a complex separable Hilbert space $\mathcal{H}$. For $w \in \mathcal{H}$ we shall write $H(w)$ to denote the closed subspace generated by $E(B)w$, $B \in \mathcal{B}$, and $d(u, H(w))$ to denote the distance of a vector u in $\mathcal{H}$ from $H(w)$.

4.3 Theorem. *If the spectral measure E has multiplicity $\geq N$, then there exist N unit vectors $u_1, u_2, \ldots, u_N \in \mathcal{H}$ such that for any $w \in \mathcal{H}$,*

$$\sum_{i=1}^{N} d^2(u_i, H(w)) \geq N - 1.$$

Proof. In light of the Hahn-Hellinger theorem in its first form we can assume that $\mathcal{H}$ is the direct sum of Hilbert spaces $L^2(X, \mathcal{B}, \mu \mid_{S_n})$, where μ is a probability measure on $\mathcal{B}$ and S_n, $n \in \mathbb{N}$, are sets in $\mathcal{B}$ such that for all n, $S_n \supseteq S_{n+1}$. The spectral measure E acts on this Hilbert space as follows: $E(A)f = 1_A f$, $f \in \mathcal{H}$. The multiplicity of E is equal to the first integer n for which $\mu(S_{n+1}) = 0$. If there is no such n the multiplicity of E is $\aleph_0$.

Now suppose that the multiplicity of $E \geq N$. Let $u_1, u_2, \ldots, u_N$ be the following unit vectors in $\mathcal{H}$:

$$u_k = (0, \ldots, 0, (\mu(S_N))^{-1/2} 1_{S_N}, 0, \ldots), \quad k = 1, 2, \ldots, N,$$

where the non-zero term appears in the kth place. If $w = (f_1, f_2, \ldots)$ is a vector in $\mathcal{H}$ then the collection of vectors of the form $(f_1 \phi, f_2 \phi, \ldots)$, where ϕ runs over bounded functions in $L^2(X, \mathcal{B}, \mu)$, is dense in $H(w)$. Let

$$v_k = \frac{1}{\sqrt{\mu(S_N)}}(f_1\phi_k, f_2\phi_k, \ldots), \quad k = 1, 2, \ldots, N,$$

be N such vectors in $H(w)$. Now

$$\sum_{i=1}^{N} d^2(u_i, H(w)) = \inf \sum_{i=1}^{N} \| u_i - v_i \|^2 \tag{1}$$

(where the infimum is taken over all choices of $v_1, v_2, \ldots, v_N$ of the given form).
Consider

$$\sum_{i=1}^{N} \| u_i - v_i \|^2 \geq (\mu(S_N))^{-1} \int_{S_N} \left(\sum_{i=1}^{N} | 1 - f_i\phi_i |^2 + \sum_{i\neq j} | f_i\phi_j |^2 \right) d\mu.$$

The integrand is bigger than or equal to

$$N - 2 \sum_{i=1}^{N} | f_i \| \phi_i | + \sum_{j=1}^{N}\sum_{i=1}^{N} | f_i |^2 | \phi_j |^2,$$

which, by applying the Schwarz inequality to the middle term, can be seen to
be

$$\geq N - 1 + \left(1 - \left(\sum_{i=1}^{N} | f_i |^2 \right)^{1/2} \left(\sum_{i=1}^{N} | \phi_i |^2 \right)^{1/2} \right)^2.$$

Thus $\sum_{i=1}^{N} \| u_i - v_i \|^2 \geq N - 1$. Taking the infimum over all possible choices
of $v_1, v_2, \ldots, v_N$, the theorem follows from (1).

Approximation with Multiplicity N

4.4 Definition. A unitary operator U is said to admit a simple approximation
with multiplicity N if there exist, for each positive integer n, N collections of
vectors $\{\phi_j^i(n) : -p_j(n) \leq i \leq q_j(n)\}$, $j = 1, 2, \ldots, N$, such that:

(i) $\phi_j^i(n) \perp \phi_k^l(n)$ *if either* $j \neq k$ *or* $i \neq l$,
(ii) $U^i\phi_j^0(n) = \phi_j^i(n)$, $\quad -p_j(n) \leq i \leq q_j(n), j = 1, \ldots, N$,
(iii) for any x in the Hilbert space

$$\left\| x - \sum_{j=1}^{N} P_j(n)x \right\| \to 0 \quad as \ \ n \to 0,$$

where $P_j(n)$ denotes the orthogonal projection on the linear span of the
jth collection: $\{\phi_j^i(n) : -p_j(n) \leq i \leq q_j(n)\}$.

4.5. Let us recall that if E is the spectral measure of a unitary operator U and if w is a vector in the Hilbert space, then the closed linear span of $U^n w, n \in \mathbb{Z}$, is identical to $H(w) =$ the closed linear span of $E(B)w$, where B runs over Borel sets in the circle group. Furthermore, the spectral multiplicity of U is equal to the multiplicity of the associated spectral measure E.

4.6. Theorem. *If a unitary operator U admits a simple approximation with multiplicity N, then the spectral multiplicity of U cannot exceed N.*

Proof. Assume, in order to arrive at a contradiction, that the spectral multiplicity of U exceeds N. Then by **4.3** there exist $N+1$ unit vectors $u_1, \ldots, u_{N+1}$ such that for any $w \in \mathcal{H}$

$$\sum_{i=1}^{N+1} d^2(u_i, H(w)) \geq N. \tag{2}$$

Let $w_j(n) = \phi_j^0(n)$. Write $u_{ij}(n) = P_j(n)u_i$, so that

$$u_i = u_{i1}(n) + u_{i2}(n) + \ldots + u_{iN}(n) + h_i(n),$$

where $\| h_i(n) \| \to 0$ as $n \to \infty$. Further we have

(i) $d^2(u_{ij}(n), H(w_j(n))) = 0, \quad j = 1, 2, .., N,$
(ii) $u_{ij}(n), \quad j = 1, 2, \ldots, N$ are orthogonal.

Since $u_{ij}(n) \in H(w_j(n))$ we have

$$d^2(u_i, H(w_j(n))) = d^2(u_{i1}(n) + \ldots + u_{iN}(n) + h_i(n), H(w_j(n)))$$

$$\leq \sum_{k=1}^{N} \| u_{ik}(n) \|^2 - \| u_{ij}(n) \|^2 + \| h_i(n) \|^2 .$$

Summing over i and using (2) we have

$$N \leq \sum_{i=1}^{N+1} \left(\sum_{k=1}^{N} \| u_{ik}(n) \|^2 - \| u_{ij}(n) \|^2 \right) + \sum_{i=1}^{N} \| h_i(n) \|^2 .$$

Summing over j,

$$N^2 \leq (N-1) \sum_{i=1}^{N+1} \sum_{k=1}^{N} \| u_{ik}(n) \|^2 + N \sum_{i=1}^{N+1} \| h_i(n) \|^2 .$$

Letting $n \to \infty$ we get $N^2 \leq (N-1)(N+1)$ which is a contradiction. Hence the multiplicity of U cannot exceed N and the proof is over.

Rank and Multiplicity

4.7. The rank of a measure preserving automorphism is related to the multiplicity of the associated unitary operator. More generally, the rank of a non-singular automorphism in a certain class bounds the spectral multiplicity of the associated unitary operator. There is an open problem in this connection which we will mention in the sequel.

4.8 Definition. Let $(X, \mathcal{B}, m)$ be a probability space. A pairwise disjoint collection $\xi = (C_0, C_1, \ldots, C_q)$ of sets in $\mathcal{B}$ is called a partition. (Here it is not required that the union of C_i's be X). For $A \in \mathcal{B}$ we write $A(\xi)$ to mean any set which is a union of sets in ξ with $m(A \Delta A(\xi))$ minimum. A sequence $\xi(n)$, $n \in \mathbb{N}$, of partitions is said to converge to the unit partition, and we write $\xi(n) \to \epsilon$, provided that for each measurable set A, $m(A \Delta A(\xi(n))) \to 0$ as $n \to \infty$.

4.9 Definition. A non-singular automorphism σ on $(X, \mathcal{B}, m)$ is said to admit a simple approximation with multiplicity N if there exists a sequence $\xi(n)$, $n \in \mathbb{N}$, of partitions such that the sets in $\xi(n)$ may be indexed as follows:

$$\xi(n) = \{C_j^i(n) : i = 1, 2, \ldots, s_j(n), \quad j = 1, 2, \ldots, N\},$$

and we further have

   (i) $\xi(n) \to \epsilon$ *as* $n \to \infty$,
   (ii) $\sigma C_j^i(n) = C_j^{i+1}(n), i = 1, \ldots, s_j(n) - 1; \ j = 1, 2, \ldots, N.$

The sequence $\xi(n)$, $n \in \mathbb{N}$, is then said to simply approximate σ with multiplicity N.

4.10. We say that σ has rank N in the measure theoretic sense if it admits a simple approximation with multiplicity N, but not with multiplicity $N - 1$. We say that σ has measure theoretic rank $\aleph_0$ if it does not have finite rank.
In case the sets $\{C_j^i(n) : i = 1, 2, \ldots, s_j(n) - 1, \ j = 1, 2, \ldots, N, n \in \mathbb{N}\}$, generate $\mathcal{B}$, then σ is said to have rank at most N in the descriptive sense. In case σ has rank at most N but not at most $N - 1$ (in the descriptive sense) then we say that σ has rank N in the descriptive sense or descriptive rank N.

     It is easy to see that if σ has descriptive rank N then σ has measure theoretic rank N. In particular, if σ has descriptive rank N then σ admits a simple approximation with multiplicity N.

4.11 Theorem. *Let σ be a non-singular automorphism on $(X, \mathcal{B}, m)$ and let ϕ be a measurable function on X of absolute value one. Let*

$$\xi(n) = \{C_j^i(n) : 1 \leq i \leq s_j(n), j = 1, 2, \ldots, N\}$$

be a sequence of partitions simply approximating σ with multiplicity N. If ϕ and $\frac{dm_\sigma}{dm}$ are constant on each $C_j^i(n)$ for every n, then the unitary operator V_ϕ on $L^2(X, \mathcal{B}, m)$ defined by

$$V_\phi f = \phi \sqrt{\frac{dm_\sigma}{dm}} f \circ \sigma, \quad f \in L^2(X, \mathcal{B}, m),$$

has spectral multiplicity at most N.

Proof. This follows at once from theorem **4.6** since V_ϕ admits a simple approximation with multiplicity N. To see this we set $\phi_j^0(n)$ equal to the indicator function of the $([\frac{1}{2}s_j(n)] + 1)$th set in the jth column $(C_j^1(n), \ldots, C_j^{s_j(n)}(n))$. We further set $p_j(n) = q_j(n) = [\frac{1}{2}s_j(n)]$. The system

$$\{\phi_j^i(n) : -p_j(n) \leq i \leq q_j(n)\}, \quad \phi_j^i(n) = \phi_j^0(n) \circ \sigma^i; \quad j = 1, 2, \ldots, N; n \in \mathbb{N},$$

then approximates V_ϕ with multiplicity N.

4.12. It does not seem to be known whether U_σ has multiplicity one whenever σ has rank one in the measure theoretic sense, σ being now assumed to be merely non-singular rather than satisfying the stronger condition that $\frac{dm_\sigma}{dm}$ is constant on sets in $\xi(n)$, $n \in \mathbb{N}$. A particular case of interest is when σ is the odometer on $X = \{0, 1\}^{\mathbb{N}}$ with measure on X given by the Markov chain with transition probability matrix $\begin{pmatrix} p & q \\ q & p \end{pmatrix}$, $p + q = 1$, $p \neq q$ and stationary initial distribution $(1/2, 1/2)$. More generally, for a non-singular σ, it is not known if U_σ has spectral multiplicity $\leq N$ whenever σ has descriptive rank N.

4.13. It is natural to ask whether σ has a simple approximation with multiplicity N whenever U_σ has spectral multiplicity N. However A. del Junco [2] has shown that U_σ may have spectral multiplicity one without σ having rank one. Indeed his example of σ is a skew product which admits a simple approximation with multiplicity two, but not one, and U_σ has spectral multiplicity one.

4.14. In the Hahn-Hellinger theorem in its second formulation the invariants of a spectral measure E are given by a sequence of mutually singular measure classes $[\mu_\infty], [\mu_1], [\mu_2], \ldots$, (some of which may be zero) such that, for each i, $[\mu_i]$ occurs with uniform multiplicity i. Write A_E for the set of those i for which $[\mu_i]$ is non-zero. It is now known as a culmination of works of E. A. Robinson Jr [6], [7], G. R. Goodson, J. Kwiatkowski, M. Lemańczyk, P. Liardet [3], J. Kwiatkowski and M. Lemańczyk [5] that given any set A of positive integers containing 1 there is a measure preserving automorphism σ such that $A = A_{E_\sigma}$ where E_σ denotes the spectral measure of U_σ. Moreover σ can be chosen to be weakly mixing. (See: M. Lemańczyk "Introduction to Ergodic Theory from the Point of View of Spectral Theory $\cdots$", Lecture Notes on the Tenth Kaisk Mathematics Workshop, Geon Ho Choe (ed), Korea Advanced Institute of Science and Technology, Math. Res. Center, Taejon, Korea.)

Chapter 5

The Skew Product

The Skew Product: Definition and its Measure Preserving Property

5.1. In this chapter we will discuss the idea of skew product in its 'simplest' form. It is a very useful method, due to Anzai, of constructing new automorphisms from known ones.

5.2. Let τ be a measure preserving automorphism on a probability space $(X, \mathcal{B}, m)$. Let ϕ be a measurable function on X with values in the circle group S^1 or a closed subgroup of S^1. Let C denote S^1 or a closed subgroup of S^1, containing the range of ϕ. Let C be given its normalised Haar measure and denote it by h. Let Y denote the cartesian product $X \times C$ equipped with its product Borel structure and product measure $m \times h$. On Y define a new automorphism T as follows:

$$T(x, c) = (\tau x, \phi(x)c);$$

T is called the skew product of τ and ϕ.

5.3. The skew product T preserves the measure $m \times h$ on Y. We see this as follows: Let $A \subseteq Y$ be a measurable set. Let $(A)_x$ denote the x-section of A. Now

$$m \times h(TA) = \int \int_{X \times C} 1_{TA}(x, c) d(m \times h) = \int_X h((TA)_x) dm(x).$$

Now $(x, c) \in TA$ if and only if $(\tau^{-1}x, \overline{\phi}(\tau^{-1}x)c) \in A$, where $\overline{\phi}$ is the complex conjugate of ϕ. Hence

$$c \in (TA)_x \Leftrightarrow \overline{\phi}(\tau^{-1}x)c \in (A)_{\tau^{-1}x}.$$

Since h is a Haar measure, we have

$$h((TA)_x) = h\big(\phi(\tau^{-1}x)(A)_{\tau^{-1}x}\big) = h\big((A)_{\tau^{-1}x}\big)$$

Thus

$$m \times h(TA) \;=\; \int_X h\big((TA)_x\big)\,dm(x)$$

$$=\; \int_X h\big((A)_{\tau^{-1}x}\big)\,dm(x)$$

$$=\; \int_X h\big((A)_x\big)\,dm(x) \;\;\text{(by the invariance of } m \text{ under } \tau)$$

$$=\; m \times h(A).$$

Thus $m \times h$ is T-invariant. We will show in the rest of this chapter that the spectrum of U_T is completely described in terms of the spectra of U_τ and V_{ϕ^n}, $n \in \mathbb{Z}$, where V_ϕ denotes the unitary operator

$$(V_\phi f)(x) = \phi(x)f(\tau x), \quad f \in L^2(X, \mathcal{B}, m).$$

The Skew Product: Its Spectrum

5.4. Consider $L^2(X \times C, m \times h)$. Fix an f in this space. For a.e. $x \in X$, $f(x, .)$ belongs to $L^2(C, h)$ and we can write

$$f(x, c) = \sum a_n(x)\chi_n(c) = \sum a_n(x)c^n, \quad \sum \mid a_n(x) \mid^2 \in L^1(X, \mathcal{B}, m),$$

where the sum is taken over all continuous characters χ_n of C. Note that the character group of C has N elements if C has N elements. If $C = S^1$ then the character group is $\mathbb{Z}$. In any case $\chi_n(c) = c^n$. Conversely, if the a_n's are measurable functions on X such that

$$\sum \mid a_n(x) \mid^2 \in L^1(X, \mathcal{B}, m),$$

then

$$f(x, c) = \sum a_n(x)c^n \;\text{ is in } L^2(X \times C, m \times h).$$

5.5. Let $\mathcal{H}_n$ be the subspace of $L^2(X \times C, m \times h)$ of functions of the form

$$f_n(x, c) = a_n(x)\chi_n(c),$$

where $a_n \in L^2(X, \mathcal{B}, m)$. We can denote $\mathcal{H}_n$ by $\chi_n L^2(X, \mathcal{B}, m)$. It is clear that $\mathcal{H}_n$ and $\mathcal{H}_k$ are orthogonal whenever $n \neq k$, because χ_n and χ_k are orthogonal. Further $\mathcal{H} = \sum_{n=-\infty}^{\infty} \mathcal{H}_n$. Now each $\mathcal{H}_n$ is invariant under U_T because if $a_n\chi_n \in \mathcal{H}_n$, then

$$(a_n\chi_n)oT(x, c) = a_n(\tau x)\chi_n(\phi(x)c) = a_n(\tau x)(\phi(x))^n c^n,$$

which is again of the same form. Now the map

$$S_n : L^2(X, \mathcal{B}, m) \to \mathcal{H}_n, \quad S_n f = \chi_n f, \quad f \in L^2(X, \mathcal{B}, m),$$

is an invertible isometry satisfying

$$(S_n^{-1} U_T S_n f)(x) = (\phi(x))^n f(Tx).$$

This shows that the restriction of U_T to $\mathcal{H}_n$ is unitarily equivalent to V_{ϕ^n} acting on $L^2(X, \mathcal{B}, m)$. The spectrum of U_T is therefore determined by the spectra of V_{ϕ^n}, $n \in \mathbb{Z}$, $(V_{\phi^0} = U_\tau)$.

5.6. When $C = \{-1, +1\}$, then

$$L^2(X \times C, m \times h) = L^2(X, \mathcal{B}, m) + \chi_1 L^2(X, \mathcal{B}, m),$$

where χ_1 is the non-trivial character on C, $\chi(1) = 1$, $\chi_1(-1) = -1$. The maximal spectral type of U_T is the sum of the maximal spectral types of U_τ and V_ϕ.

5.7. In **5.1** we used the adjective 'simplest' to describe the skew product discussed above, because τ is assumed to preserve a finite measure and ϕ is assumed to take values in the circle group. Now one can assume that τ is non-singular and that ϕ takes values in a locally compact group G. The skew product T is then defined on $X \times G$. It is even necessary to discuss such T which arise when G is a Polish group, such as the group of unitary operators on a Hilbert space. The case when G is locally compact is discussed in K. Schmidt [2]. The spectrum of such U_T when G is non-Abelian is discussed in E. Robinson, Jr. [3].

Chapter 6

A Theorem of Helson and Parry

Statement of the Theorem

6.1. Let τ be a measure preserving aperiodic automorphism on a standard probability space $(X, \mathcal{B}, m)$. Let ϕ be a Borel function on X of absolute value one and consider the unitary operator V_ϕ defined on $L^2(X, \mathcal{B}, m)$ by

$$(V_\phi f)(x) = \phi(x) f(\tau x), \quad f \in L^2(X, \mathcal{B}, m).$$

In a paper entitled "Cocycles and Spectra" [3], Helson and Parry prove that for every aperiodic τ there exists a function ϕ such that the maximal spectral type of V_ϕ is Lebesgue; moreover ϕ can be chosen to be real, i.e., taking values -1 and $+1$. The purpose of this chapter is to prove a version of this theorem for hyperfinite actions of countable groups. The improved version is obtained by combining the method of Helson and Parry with the notion of orbit equivalence.

Weak von Neumann Automorphisms and Hyperfinite Actions

6.2. We recall the definition of a weak von-Neumann automorphism. A non-singular automorphism τ on $(X, \mathcal{B}, m)$ is said to be a weak von Neumann if there exist a sequence

$$\mathcal{D}_k(\tau) = (D_1^k, D_2^k, \ldots, D_{2^k}^k), \quad k \in \mathbb{N},$$

of ordered partitions of X into measurable sets such that

(a) $D_i^k = D_i^{k+1} \cup D_{i+2^k}^{k+1}$, $i = 1, 2, \ldots, 2^k$; $k = 1, 2, \ldots$,
(b) $D_i^k = \tau^{i-1} D_1^k$, $i = 1, 2, \ldots, 2^k$; $k = 1, 2, \ldots$.

If the sets D_i^k, $i = 1, 2, \ldots, 2^k$; $k \in \mathbb{N}$, generate the σ-algebra $\mathcal{B}$, then τ is said to be a von Neumann automorphism. We will not need von Neumann automorphisms in this chapter.

6.3. Write $orb(x, n)$ to mean the finite set $(x, \tau x, \tau^2 x, \ldots, \tau^{n-1} x)$ if $n \geq 0$, and the set $(\tau^{-1} x, \tau^{-2} x, \ldots, \tau^{-n} x)$ if $n < 0$. If we put

$$F_k = D_{2^k-1}^k, \quad F_{k,1} = D_{2^k-1}^{k+1}, \quad F_{k,2} = D_{2^k-1+2^k}^{k+1},$$

then $F_k = F_{k,1} \cup F_{k,2}$ and for $2^k \leq n < 2^{k+1}$, $orb(x, n)$ intersects F_k in at least one point and intersects $F_{k,1}$ and $F_{k,2}$ in at most one point. This fact will be needed later.

6.4. Let G be a countable group, not necessarily Abelian, but written additively. Let $T_g, g \in G$, be a group of non-singular automorphisms on X weakly equivalent to a weak von Neumann automorphism. This means that there is a weak von Neumann automorphism τ on X such that for a.e. x, the orbit of x under τ is the same as the orbit of x under T_g, $g \in G$. Such an action of a group G is called hyperfinite. It is known from a theorem of Connes, Feldman and Weiss [2] that if G is countable and amenable then any non-singular action of G is hyperfinite and the converse holds. In particular, measure preserving actions of countable Abelian groups are hyperfinite.

For simplicity we will assume in the rest of this chapter that the action $T_g, g \in G$, and the weak von Neumann automorphism τ are measure preserving. Further $T_g, g \in G$, will be assumed to be free (i.e., for any x, $T_g x = x$ only if $g = e$ the identity element of G) and weakly equivalent to the weak von Neumann automorphism τ.

The Cocycle $C(g, x)$

6.5. Let $T_g, g \in G$, and τ be as above and define the function C on $G \times X \to \mathbb{Z}$ by

$$C(g, x) = n \text{ if } T_g x = \tau^n x.$$

For fixed g and n the set $\{x : C(g, x) = n\} = \{x : T_g x = \tau^n x\}$ is measurable. Moreover it can be verified that C satisfies the cocycle identity

$$C(h + g, x) = C(g, x) + C(h, T_g x) \quad a.e. \text{ for all } g, h \in G.$$

6.6 Lemma. *Given $\varepsilon > 0$ and a positive integer k there exists a finite set $S \subset G$ such that if*

$$Q(k, g) = \{x : \ |C(g, x)| < 2^k\}$$

then

$$\sum_{g \notin S} m(Q(k, g)) < \varepsilon.$$

Proof. Let $g_1, g_2, g_3, \ldots$, be an enumeration of G. Then for a.e. x, $|C(g_n, x)| \to \infty$ as $n \to \infty$. By Egorov's theorem there exists a set B of m measure less than

$\frac{\varepsilon}{2^{k+1}}$ such that $\mid C(g_n, x) \mid \to \infty$ uniformly on $X - B$. Choose N so large that if $g \notin \{g_1, g_2, \ldots, g_N\} = S$ then $\mid C(g, x) \mid \geq 2^k$ for $x \in X - B$. Now

$$\sum_{g \notin S} m(Q(k, g)) = \sum_{g \notin S} \sum_{|j| < 2^k} m(\{x : C(g, x) = j\})$$

$$= \sum_{|j| < 2^k} \sum_{g \notin S} m(\{x : C(g, x) = j\}).$$

For fixed j, the sets $\{x : C(g, x) = j\}$ are all disjoint as g runs over G and for $g \notin S$ they are contained in B. Hence

$$\sum_{g \notin S} m(Q(k, g)) \leq \sum_{|j| < 2^k} m(B) < \varepsilon,$$

by choice of the set B. The lemma is proved.

6.7 Remark. If S satisfies the conclusion of the above lemma then any finite subset of G containing S also satisfies the conclusion of the lemma. In view of this, we have the following:

6.8 Corollary. *There exist finite sets $S_k \subseteq G$, $k = 0, 1, 2, 3, \ldots$, such that*

(i) $S_0 = \emptyset$, $S_k \subseteq S_{k+1}$ *for all* k,
(ii) *for* $g \in S_{k+1} - S_k$ *if we set*

$$Q(g) = \{x : \mid C(g, x) \mid < 2^k\} = Q(k, g),$$

then

$$\sum_{g \in G} m(Q(g)) < \varepsilon.$$

Proof. Choose positive ε_k, $k \in \mathbb{N}$, such that their sum is finite. For each $k \geq 1$, choose $S_k \supseteq S_{k-1}$ such that $\sum_{g \notin S_k} m(Q(k, g)) < \varepsilon_k$. This is possible by lemma **6.6** and remark **6.7**. We may assume that $\bigcup S_k = G$. Then

$$\sum_{g \in G} m(Q(g)) = \sum_{k=0}^{\infty} \sum_{g \in S_{k+1} - S_k} m(Q(k, g)) \leq \sum_{k=0}^{\infty} \sum_{g \notin S_k} m(Q(k, g)) < \infty,$$

where we have used the fact that $Q(0, g)$ is empty except for $g = 0$.

The Random Cocycle and the Main Theorem

6.9. Equipped with the above facts we now proceed to formulate the main theorem. Let p be a real measurable function on X. We shall write

$$\sum^{n} p(\tau^k x) = \begin{cases} +\sum_{k=0}^{n-1} p(\tau^k x) & \text{if } n > 0, \\ 0 & \text{if } n = 0, \\ -\sum_{k=1}^{-n} p(\tau^{-k} x) & \text{if } n < 0. \end{cases}$$

The function $\phi(n,x) = exp(i\sum^n p(\tau^k x))$ is a $\mathbb{Z} \times X$ cocycle (relative to τ) taking values in the circle group. The function

$$A(g,x) = \phi(C(g,x),x)$$

can be verified to be a $G \times X$ cocycle (relative to $T_g, g \in G$), i.e., $A(g,.)$ is a Borel function for each g and satisfies

$$A(g+h,x) = A(h,x)A(h,T_g x), \quad g, \ h \in G, \ x \in X.$$

Define the unitary group

$$(V_g f)(x) = A(g,x)f(T_g x), \ f \in L^2(X,\mathcal{B},m). \tag{2}$$

We are now ready to state the version of the theorem of Helson and Parry mentioned in **6.1** The proof follows the pattern of the first construction of their paper [3].

6.10 Theorem. *Let $T_g, g \in G$, be a measure preserving action of a countable group G on $(X,\mathcal{B},m)$ which is orbit equivalent to a weak von Neumann automorphism τ. Then there exists a $G \times X$ cocycle A taking values -1 and +1 such that for all $f \in L^\infty(X,\mathcal{B},m)$,*

$$\sum_{g \in G} |(V_g f, f)|^2 < \infty.$$

Proof.
(1) Let $C(g,x)$ denote the integer n such that $T_g x = \tau^n x$. Let

$$E_j, j = 1, 2, 3, \ldots,$$

be disjoint measurable sets in X with indicator functions

$$h_j, j = 1, 2, 3, \ldots,$$

respectively. Form the random set E whose indicator function h is

$$\sum_{j=1}^{\infty} \eta_j h_j$$

where η_j's are independent random variables taking values 0 and 1 each with probability 1/2. Define $p = \pi h$ which is now a function of x and w, where w is in the space on which η_j's are defined. Set

$$\phi(n,x) = \exp\left(i\sum^n p(\tau^k x)\right),$$

$$A(g,x) = \phi(C(g,x),x),$$

which are now $\mathbb{Z} \times X$ and $G \times X$ cocycles respectively. Fix an $f \in L^\infty(X, \mathcal{B}, m)$ and let $\rho(g) = (V_g f, f)$, where $V_g, g \in G$, is defined by (2), using the random cocycle A. Note that $\rho(g)$ depends on w. We will show that the E_j's can be so chosen that for all $f \in L^\infty(X, \mathcal{B}, m)$, $\sum_{g \in G} |\rho(g)|^2 < \infty$ for a.e. w. A routine calculation shows that

$$|\rho(g)|^2 =$$

$$\int\int_{X \times X} (*) \exp \pi i \left(\sum_{j=1}^\infty \left(\sum^{C(g,x)} \eta_j h_j(\tau^k x) - \sum^{C(g,y)} \eta_j h_j(\tau^k y) \right) \right) dm(x) dm(y),$$

where $(*)$ in the integrand stands for the expression

$$f(T_g x)\overline{f}(x)\overline{f}(T_g y)f(y).$$

Integrating over the probability space Ω on which the η_j's are defined gives

$$\int_\Omega |\rho(g)(\omega)|^2 \, d\omega =$$

$$\int\int_{X \times X} (*) \prod_{j=1}^\infty \frac{1}{2}\left[1 + \exp \pi i \left(\sum^{C(g,x)} h_j(\tau^k x) - \sum^{C(g,y)} h_j(\tau^k y) \right) \right] dm(x) dm(y).$$

$$(A)$$

(2) The product on the right hand side takes values 0 and 1 and equal to one on the set in $X \times X$ consisting of all pairs (x, y) such that

$$\text{parity of } \sum^{C(g,x)} h_j(\tau^k x) = \text{parity of } \sum^{C(g,y)} h_j(\tau^k y),$$

for all $j = 1, 2, 3, \ldots$. Write

$$a_j^g(x) = \begin{cases} 0 & \text{if } orb(x, C(g, x)) \cap E_j \text{ has even number of points,} \\ 1 & \text{if } orb(x, C(g, x)) \cap E_j \text{ has odd number of points.} \end{cases}$$

Let $a^g(x) = (a_1^g(x), a_2^g(x), a_3^g(x), \ldots)$, a sequence of zeros and ones, terminating in zeros since for each x, $orb(x, C(g, x))$ is a finite set and $E_1, E_2, E_3, \ldots$, are pairwise disjoint non-empty sets. For each sequence a of zeros and ones terminating in zeros let G_a^g be the set of $x \in X$ such that $a^g(x) = a$. For g fixed, the sets G_a^g form a disjoint covering of X as a runs over all sequences of zeros and ones terminating in zeros. Evidently $a^g(x) = a = a^g(y)$ if and only if x and y belong to G_a^g. The $m \times m$ measure of this set of (x, y) is $(m(G_a^g))^2$. Thus by (A)

$$\int_\Omega |\rho(g)|^2 \, d\omega \leq \| f \|_\infty^4 \sum_a (m(G_a^g))^2.$$

(3) Since τ is a weak von Neumann automorphism we have sets $F_{k,1}, F_{k,2}$ as per **6.3**, for $k = 1, 2, 3, \ldots$. We can decompose $F_{k,1}, F_{k,2}$ into finitely many sets

$F_k^1, F_k^2, \ldots, F_k^{l_k}$ such that for $1 \leq l \leq l_k$, the sets

$$\bigcup_{|s|<2^{k+1}} \tau^s F_k^l$$

have measure less than δ_k, where δ_k's will be chosen later. Let

$$E_j, j = 1, 2, 3, \ldots$$

be an enumeration of the sets F_k^l, $l = 1, 2, 3, \ldots$, $k \in \mathbb{N}$. These are the sets needed to prove the theorem.

(4) The E_j's are pairwise disjoint by construction and their union is all of X. Each E_j is contained a unique $F_k = F_{k,1} \cup F_{k,2}$. If $2^k \leq |n| < 2^{k+1}$, then by **6.3**, for each n, $orb(x, n)$ intersects F_k and then such an orbit intersects F_k^l in at most one point since each F_k^l is contained in only one of $F_{k,1}$ or $F_{k,2}$.

(5) Fix a g and let r be such that $g \in S_r - S_{r-1}$. The measure of $m(G_a^g)$ is to be estimated. Now

$$m(G_a^g) \leq \sum_{k=r}^{\infty} m(G_a^g \cap \{x : 2^k \leq |C(g,x)| < 2^{k+1}\}) + m(Q(g)),$$

where $Q(g)$ is as in corollary **6.8**. We now estimate the kth term of the summation. Let $a = (a_1, a_2, a_3, \ldots)$, be a sequence of zeros and ones terminating in zeros. Suppose $a_j = 0$ for each j for which $E_j \subseteq F_k$. If $2^k \leq |C(g,x)| < 2^{k+1}$, then $orb(x, C(g,x))$ intersects some $E_j \subseteq F_k$ in exactly one point. Hence $G_a^g \cap \{x : 2^k \leq |C(g,x)| < 2^{k+1}\}$ is empty if $a_j = 0$ for each j with $E_j \subseteq F_k$. For such a,

$$m(G_a^g \cap \{x : 2^k \leq |C(g,x)| < 2^{k+1}\}) = 0.$$

Otherwise $a_j = 1$ for at least one j such that $E_j \subseteq F_k$. The set

$$G_a^g \cap \{x : 2^k \leq |C(g,x)| < 2^{k+1}\}$$

is then contained in the set of all x such that $orb(x, C(g,x))$ intersects E_j and $2^k \leq |C(g,x)| < 2^{k+1}$. Thus

$$G_a^g \cap \{x : 2^k \leq |C(g,x)| < 2^{k+1}\} \subseteq \bigcup_{|s|<2^{k+1}} \tau^s E_j,$$

$$m(G_a^g \cap \{x : 2^k \leq |C(g,x)| < 2^{k+1}\}) \leq \delta_k,$$

$$m(G_a^g) \leq \sum_{k \geq r} \delta_k + m(Q(g)) = \gamma_r + m(Q(g)),$$

where we have put $\gamma_r = \sum_{k \geq r} \delta_k$.

Now for a fixed $g \in S_r - S_{r-1}$, G_a^g form a disjoint covering of X as a runs over sequences of zeros and ones terminating in zeros. Hence

$$\sum_a \left(m(G_a^g) \right)^2 \leq \left(\gamma_r + m(Q(g)) \right) \cdot \sum_a m(G_a^g) \leq \gamma_r + m(Q(g)).$$

Finally, summing over $g \in G$, we get

$$\sum_{g \in G} \sum_a \left(m(G_a^g) \right)^2 \leq \sum_{r=1}^{\infty} \mid S_r - S_{r-1} \mid \gamma_r + \sum_{g \in G} m(Q(g)), \qquad (B)$$

where $\mid S_r - S_{r-1} \mid$ = number of elements in $S_r - S_{r-1}$. We now choose the δ_k's in such a way that the first sum on the right hand side of (B) is convergent. The second sum is convergent by corollary **6.8**.

(6) Thus $\sum_{g \in G} \int_\Omega \mid \rho(g) \mid^2 d\omega < \infty$. Hence for almost every ω,

$$\sum_{g \in G} \mid (V_g f, f) \mid^2 < \infty. \qquad (C)$$

The null set of points ω where the sum may not converge depends on f. But $L^2(X, \mathcal{B}, m)$ is separable, hence there is a grand null set N of points ω such that the sum (C) converges for all $\omega \notin N$ and every $f \in L^\infty(X, \mathcal{B}, m)$. Theorem **6.10** is thus proved.

Remarks

6.11 Remark 1. Although we have proved theorem **6.10** for measure preserving hyperfinite actions, essentially the same method allows us to prove the theorem for non-singular hyperfinite actions (see J. Mathew and M. Nadkarni [5]).

Remark 2. In case the action $T_g, g \in G$, is the action of integers

$$\tau^n, n \in \mathbb{Z},$$

where τ is the measure preserving von Neumann automorphism, then ϕ can be chosen more concretely as follows: for each k define $\phi = 1$ on F_{k1} and equal to -1 on F_{k2}, F_{k1}, F_{k2} being as in section **6.3**. Then it can be shown directly that V_ϕ has Lebesgue spectrum with multiplicity 2. (See J. Mathew and M. Nadkarni [6], M. Quéffelec [7].) The method can be generalised to obtain Lebesgue spectrum of any even multiplicity. (See O. Ageev [1], M. Lemańczyk [4]).

Remark 3. In case G is a countable Abelian group, the above theorem shows that the cocycle A can be chosen so that unitary group $V_g, g \in G$, has spectral measure absolutely continuous with respect to the Haar measure on the compact dual of G.

Remark 4. When $G = \mathbb{Z}$ we can choose A so that the corresponding $V_n, n \in \mathbb{Z}$, has Lebesgue spectrum. This is done by ensuring that the series (C) converges very fast. See Helson and Parry [2].

Remark 5. If $T_g^{(1)}$, $g \in G_1, \ldots, T_g^{(n)}$, $g \in G_n$ are finitely many actions of countable groups $G_1, \ldots, G_n$, all having orbits the same as a single weak von Neumann automorphism τ on X, then there exists a single $\mathbb{Z} \times X$ cocycle ϕ relative to τ such that the associated $V_g^{(j)}$, $g \in G_j, j = 1, 2, \ldots, n$, defined in **6.10** all satisfy (C) for all $f \in L^\infty(X, \mathcal{B}, m)$. This can be accomplished by choosing δ_k's sufficiently small.

Chapter 7

Probability Measures on the Circle Group

Continuous Probability Measures on S^1: They are Dense G_δ

7.1. In this chapter we will study certain aspects of finite measures on the circle group which can be applied to the spectral theory of non-singular automorphisms.

7.2. Let $\mathcal{P}$ denote the space of all probability measures on the Borel σ-algebra of the circle group S^1. Let $\mathcal{P}_C$ denote the subset of $\mathcal{P}$ of measures which have no point masses. For $\mu \in \mathcal{P}$, and $k \in \mathbb{Z}$, $\hat{\mu}(k)$ will denote the kth Fourier co-efficient of μ :

$$\hat{\mu}(k) = \int_{S^1} z^{-k} d\mu.$$

7.3. The sequence $\hat{\mu}(k)$, $k \in \mathbb{Z}$, uniquely determines μ. Since $\mid \hat{\mu}(k) \mid \leq 1$, the map $\mu \to (\hat{\mu}(k))_{k \in \mathbb{Z}}$ is a one-one map of $\mathcal{P}$ into $\prod_{k=-\infty}^{\infty} D_k$ where each D_k is a copy of the closed unit disk D. The image J of this map is a closed subset of $\prod_{k=-\infty}^{\infty} D_k$ (equipped with its product topology). We see this as follows:

Suppose μ_n, $n \in \mathbb{N}$, is a sequence in $\mathcal{P}$ and $(c_n), n \in \mathbb{Z}$, is an element of $\prod_{k=-\infty}^{\infty} D_k$ such that $\hat{\mu}_n(k) \to c_k$ for each k, then the sequence (c_k), $k \in \mathbb{Z}$, is positive definite with $c_0 = 1$. By Herglotz' theorem there exists a $\mu \in \mathcal{P}$ such that for each k, $\hat{\mu}(k) = c_k$. Thus J is a closed subset of $\prod_{k=-\infty}^{\infty} D_k$. We lift the compact metric topology of J to $\mathcal{P}$; it is the weak (or vague) topology on $\mathcal{P}$ in which

$$\mu_n \to \mu \Leftrightarrow \int_{S^1} f d\mu_n \to \int_{S^1} f d\mu,$$

for all continuous f on S^1.

7.4. For fixed $k \in \mathbb{Z}$, the function

$$\mu \to \hat{\mu}(k)$$

is continuous. Hence for each n the function

$$\mu \to \frac{1}{2n+1} \sum_{k=-n}^{n} |\hat{\mu}(k)|^2$$

is continuous. By Wiener's Lemma

$$\lim_{n \to \infty} \frac{1}{2n+1} \sum_{k=-n}^{n} |\hat{\mu}(k)|^2 = \sum_{x \in S^1} |\mu(x)|^2 . \qquad (1)$$

Hence $\mu \in \mathcal{P}_C$ if and only if

$$\lim_{n \to \infty} \frac{1}{2n+1} \sum_{k=-n}^{n} |\hat{\mu}(k)|^2 = 0.$$

7.5. Since the limit $\lim_{n \to \infty} \frac{1}{2n+1} \sum_{k=-n}^{n} |\hat{\mu}(k)|^2$ exists we see from (1) that

$$\mu \in \mathcal{P}_C \Leftrightarrow \liminf_{n \to \infty} \frac{1}{2n+1} \sum_{k=-n}^{n} |\hat{\mu}(k)|^2 = 0.$$

Equivalently,

$$\mu \in \mathcal{P}_C \Leftrightarrow \forall \ m \ \forall \ l \ \exists \ n \geq l \text{ such that } \frac{1}{2n+1} \sum_{k=-n}^{n} |\hat{\mu}(k)|^2 < \frac{1}{m}.$$

Or

$$\mu \in \mathcal{P}_C \Leftrightarrow \mu \in \bigcap_{m=1}^{\infty} \bigcap_{l=1}^{\infty} \bigcup_{n=l}^{\infty} \left\{ \nu \in \mathcal{P} : \frac{1}{2n+1} \sum_{k=-n}^{n} |\hat{\mu}(k)|^2 < \frac{1}{m} \right\}.$$

Each set in the curly bracket is open in $\mathcal{P}$ by our comment in **7.4** so that $\mathcal{P}_C$ is a G_δ set in $\mathcal{P}$. Since the set of measures in $\mathcal{P}$ absolutely continuous with respect to Lebesgue measure is dense in $\mathcal{P}$, we get the well known fact that $\mathcal{P}_C$ is dense G_δ in $\mathcal{P}$. We state this as:

7.6 Theorem. *The class of continuous measures in $\mathcal{P}$ is dense G_δ in $\mathcal{P}$, hence residual in $\mathcal{P}$.*

Remark. Our proof of theorem **7.6**, although an interesting application of Wiener's lemma, is defective in the sense that it does not generalise to the space of probability measures on a compact metric space where the theorem is still valid.

7.7 Corollary. *The set*

$$\{\mu \in \mathcal{P}_C : closed\ support\ of\ \mu\ is\ S^1\}$$

is dense G_δ *in* $\mathcal{P}_C$, *hence residual in* $\mathcal{P}_C$ *and* $\mathcal{P}$.

Proof. The class of measures in $\mathcal{P}$ which vanish on a fixed open arc $I \subseteq S^1$ is closed nowhere dense in $\mathcal{P}$. Hence the set of measures in $\mathcal{P}$ which vanish on some arc in S^1 is meager F_σ in $\mathcal{P}$.

Measures Orthogonal to a Given Measure

7.8. Two measures μ and ν in $\mathcal{P}$ are said to be orthogonal if they are mutually singular, in which case we write $\mu \perp \nu$. We need the fact: μ and ν are orthogonal if and only if for all $\varepsilon > 0$ there exists an open set U_ε such that $\mu(U_\varepsilon) > 1 - \varepsilon$ and $\nu(U_\varepsilon) < \varepsilon$. We also need the fact that for any open U the map $\mu \to \mu(U)$ is continuous on $\mathcal{P}_C$.

7.9 Theorem. *Fix a* $\nu \in \mathcal{P}$. *The set*

$$\nu^\perp = \{\mu \in \mathcal{P}_C : \mu \perp \nu\}$$

is dense G_δ *in* $\mathcal{P}_C$. *The corresponding set in* $\mathcal{P}$, *i.e., the set* $\{\mu \in \mathcal{P} : \mu \perp \nu\}$ *is residual in* $\mathcal{P}$.

Proof. Let $\mathcal{U}$ be the collection of open sets in S^1. In the light of **7.8** we have:

$$\mu \in \mathcal{P}_C \text{ and } \in \nu^\perp \Leftrightarrow$$

$$\mu \in \mathcal{P}_C \text{ and } \forall\ m \in \mathbb{N},\ \exists\ U \in \mathcal{U} \text{ with } \nu(U) < \frac{1}{m},\ \mu(U) > 1 - \frac{1}{m} \Leftrightarrow$$

$$\mu \in \bigcap_{m=1}^{\infty} \bigcup_{U \in \mathcal{U}} \{\mu \in \mathcal{P}_C : \nu(U) < \frac{1}{m},\ \mu(U) > 1 - \frac{1}{m}\}.$$

so that $\nu^\perp$ is a G_δ set in $\mathcal{P}_C$ and $\mathcal{P}$. Since there exist measures in $\mathcal{P}_C$ orthogonal to ν whose closed support is S^1 and since the collection of such measures is dense in $\mathcal{P}_C$, we see that $\nu^\perp$ is dense G_δ in $\mathcal{P}_C$. Since $\nu^\perp \subseteq \{\mu \in \mathcal{P} : \mu \perp \nu\}$ the latter set is residual in $\mathcal{P}$.

7.10 Corollary. *Given* $\nu \in \mathcal{P}$, *the set* $\{\mu \in \mathcal{P} : \mu \ll \nu\}$ *is meager in* $\mathcal{P}$. *(Here* $\mu \ll \nu$ *means* μ *is absolutely continuous with respect to* ν).

Measures Singular under Convolution and Folding

7.11. For $p \in \mathbb{N}$, let $\psi_p : S^1 \to S^1$ be the map $\psi_p(z) = z^p$. For any $\mu \in \mathcal{P}$, let $\mu^{(p)}$ and μ^p denote the measures $\mu \circ \psi_p^{-1}$ and the p-fold convolution $\mu * \mu * \cdots * \mu$ (p times) of μ with itself. For $k \in \mathbb{Z}$

$$(\mu^{(p)})^\wedge(k) = \hat{\mu}(kp), \quad (\mu^p)^\wedge(k) = (\hat{\mu}(k))^p.$$

Hence the functions $\mu \to \mu^{(p)}$ and $\mu \to \mu^p$ are continuous from $\mathcal{P}$ to $\mathcal{P}$. These facts together with the method of **7.9** permit one to prove that the sets

$$R = \{\mu \in \mathcal{P}_C : \mu^{(p)} \perp \mu^{(q)}, \quad p \neq q, \quad p, q \in \mathbb{N}\},$$

$$S = \{\mu \in \mathcal{P}_C : \mu^p \perp \mu^q, \quad p \neq q, \quad p, q \in \mathbb{N}\},$$

are G_δ. We briefly indicate how R may be shown to be G_δ. Indeed, for $\mu \in \mathcal{P}_C$ we have $\mu \in R$

$$\Leftrightarrow \forall m \in \mathbb{N} \; \forall \, p < q \; \exists \, U \in \mathcal{U} \text{ with } \mu^{(p)}(U) < \frac{1}{m}, \; \mu^{(q)}(U) > 1 - \frac{1}{m}$$

$$\Leftrightarrow \mu \in \bigcap_{m=1}^\infty \bigcap_{p<q} \bigcup_{U \in \mathcal{U}} \left\{\mu \in \mathcal{P}_C : \mu^{(p)}(U) < \frac{1}{m}, \mu^{(q)}(U) > 1 - \frac{1}{m}\right\},$$

which shows that R is a G_δ set in $\mathcal{P}_C$.

7.12. The sets R and S are dense in $\mathcal{P}_C$ as seen below:

If μ is in R ($\mu \in S$) and a ν in $\mathcal{P}$ satisfies $\nu \ll \mu$, then $\nu \in R$ ($\nu \in S$). Hence if we show that there is a $\mu \in R$ ($\mu \in S$) whose closed support is all of S^1, then it will follow that R (S) is dense in $\mathcal{P}_C$. Let $B \subseteq [0, 1]$ be an uncountable Borel set which is independent over the field of rationals and contains a rational (see Kahane and Salem [4] p 20). Let $C = \{e^{2\pi i x} : x \in B\}$ and $D = \bigcup_{r \in Q} rC$, where Q stands for elements in S^1 of the form $e^{2\pi i \frac{m}{n}}$. Now if $p \neq q$, then $\psi_p^{-1}(D) \cap \psi_q^{-1}(D) = Q$, $D^p \cap D^q = Q$, where D^p denotes the set $\{z_1 \cdot z_2 \cdots z_p : z_1, z_2, \ldots, z_p \in D\}$. Let μ be a measure in $\mathcal{P}_C$ carried by D and whose closed support is S^1. Since for all p, $\mu^{(p)}$ is carried by $\psi_p^{-1}(D)$ (μ^p carried by D^p) and $\mu^{(p)}$ is continuous (μ^p is continuous), we conclude $\mu^{(p)}$, $p \in \mathbb{N}$ (μ^p, $p \in \mathbb{N}$), are all mutually singular so that $\mu \in R$ ($\mu \in S$). Thus we have:

7.13 Theorem. *The sets R and S are dense G_δ in $\mathcal{P}_C$. The corresponding sets in $\mathcal{P}$ are residual in $\mathcal{P}$.*

If D is as in **7.12** and $\tilde{D} = \bigcup_{n=1}^\infty D^n$, then for $p, q \in \mathbb{N}$, $p \neq q$, $\psi_p^{-1}(\tilde{D}) \cap \psi_q^{-1}(\tilde{D})$ is countable. So if $\mu \in \mathcal{P}_C$ is carried by D, then $\overline{\mu} = \sum_{n=1}^\infty \frac{1}{2^n} \mu^n$ is in $\mathcal{P}_C$ and carried by $\tilde{D}$. Further $\overline{\mu}^{(p)} \perp \overline{\mu}^{(q)}$ whenever $p \neq q$ so that $\overline{\mu} \in R$. More generally if μ is carried by $D \cup D^{-1}$, then $\sum_{n=1}^\infty \frac{1}{2^n} \mu^n$ is in R. We will need this fact later.

Rigid Measures

7.14. We now discuss classes of measures called rigid measures and α-rigid measures ($\mid \alpha \mid = 1$).

7.15. A measure μ in $\mathcal{P}$ is said to be rigid if there exists a sequence $(n_k), k \in \mathbb{N}$, of natural numbers such that $\hat{\mu}(n_k) \to 1$ as $k \to \infty$. More generally, for $\alpha \in S^1$, μ is said to be α-rigid if for some sequence $(n_k), k \in \mathbb{N}$, of natural numbers $\hat{\mu}(n_k) \to \alpha$ as $k \to \infty$. Any sequence $(n_k), k \in \mathbb{N}$, along which such convergence occurs is called a sequence associated with μ. If $z^{-n_k} \to \alpha$ in μ measure then clearly μ is α-rigid with associated sequence $(n_k), k \in \mathbb{N}$. Conversely, if μ is α-rigid with associated sequence $(n_k), k \in \mathbb{N}$, then $z^{-n_k} \to \alpha$ in μ-measure as seen below:

If μ is α-rigid with associated sequence $(n_k), k \in \mathbb{N}$, then $\alpha^{-1}\hat{\mu}(n_k) \to 1$ as $k \to \infty$, so that

$$\int_{S^1} [1 - Re(\alpha^{-1}z^{-n_k})]d\mu \to 0 \quad \text{as} \quad k \to \infty.$$

Since $0 \leq 1 - Re(\alpha^{-1}z^{-n_k}) \leq 2$, $1 - Re(\alpha^{-1}z^{-n_k}) \to 0$ in μ-measure as $k \to \infty$, i.e., $z^{-n_k} \to \alpha$ in μ-measure. We have proved:

7.16 Proposition. *A measure $\mu \in \mathcal{P}$ is α-rigid with an associated sequence $(n_k), k \in \mathbb{N}$, if and only if $z^{-n_k} \to \alpha$ in μ-measure as $k \to \infty$.*

7.17. It is clear from this reformulation that if μ is α-rigid with associated sequence (n_k), $k \in \mathbb{N}$, then every $\nu \in \mathcal{P}$, $\nu \ll \mu$, is also α-rigid with the same associated sequence. We further observe:

(1) A discrete probability measures is rigid because $\hat{\mu}(n)$, $n \in \mathbb{Z}$, is then almost periodic with $\hat{\mu}(0) = 1$ so that $\hat{\mu}(n) \to 1$ over some subsequence of natural numbers.

(2) If μ is α-rigid with an associated sequence $n_k, k \in \mathbb{N}$, then for all natural numbers p, μ is α^p-rigid with the associated sequence (pn_k), $k \in \mathbb{N}$. This is because $z^{-n} \to \alpha^p$ in μ-measure over the subsequence $(pn_k), k \in \mathbb{N}$. If μ is α-rigid with $\alpha = e^{2\pi i a}$, a irrational, then it follows by a diagonal argument that μ is β-rigid for every β in S^1. Thus if μ is α-rigid then μ is β-rigid for every β in the closed subgroup generated by α.

(3) If μ is α-rigid with an associated sequence $n_k, k \in \mathbb{N}$, then for every $p \in \mathbb{N}$, $\mu^{(p)}$ and μ^p are α^p-rigid with the same associated sequence.

(4) If μ is α-rigid and if $\alpha \neq 1$, then μ and $\mu * \mu$ (also μ and $\mu^{(2)}$) are mutually singular. For if λ is any member of $\mathcal{P}$ absolutely continuous with respect to μ and $\mu * \mu$ then λ is both α-rigid and α^2-rigid with the same associated sequence which is not possible unless $\alpha = 1$. A similar argument proves the mutual singularity of μ and $\mu^{(2)}$. If μ is α-rigid with $\alpha = e^{2\pi i a}$, a irrational, then $\mu, \mu^2, \mu^3, \ldots$, are all mutually singular and so are $\mu, \mu^{(2)}, \mu^{(3)}, \ldots$.

(5) If μ is α-rigid with an associated sequence $(n_k), k \in \mathbb{N}$, then $\tilde{\mu}$ defined by $\mu(A) = \mu(A^{-1})$ is α^{-1} rigid with the same associated sequence. If μ is weakly symmetric in the sense that $\mu(A) > 0$ whenever $\mu(A^{-1}) > 0$, then μ cannot be α-rigid unless $\alpha = 1$ or $\alpha = -1$. If μ is α-rigid then $\mu * \tilde{\mu}$ is 1-rigid.

(6) For any $\mu \in \mathcal{P}$ the set of limit points of the sequence $\hat{\mu}(n)$, $n \in \mathbb{N}$, which lie in S^1 form a closed subgroup of S^1.

(7) If μ is rigid then over some subsequence (n_k), $k \in \mathbb{N}$, $z^{-n_k} \to 1$ in μ measure. Over a further subsequence of (n_k), say (l_k), $k \in \mathbb{N}, z^{-l_k} \to 1$ *a.e.* μ. Hence μ is carried by the set

$$D(l_1, l_2, l_3, \dots) = \{z : z^{-l_n} \to 1, \quad \text{as } n \to \infty\}.$$

(8) For any sequence $n_1 < n_2 < n_3 \dots$, of natural numbers the set $D = D(n_1, n_2, n_3, \dots)$, is well defined and we call it the Dirichlet subgroup associated to the sequence $(n_k), k \in \mathbb{N}$. It is a subgroup of S^1 and a Borel set. Every probability measure carried by D is rigid with associated sequence $n_1, n_2, n_3, \dots$. By the Riemann-Lebesgue Lemma the set D therefore has Lebesgue measure zero. With ψ_p as in **7.11** we note that the sets $\psi_p^{-1}(D)$, $p \in \mathbb{N}$, have Lebesgue measure zero, so their union over all p also has Lebesgue measure zero. If $z \in S^1$ and $z^{n_k}, k \in \mathbb{N}$, has only a finite set of limit points of the form $exp(2\pi i \frac{p}{q})$, $(p, q$ integers, $q > 0)$, then for some integer p_0, $z^{-p_0 n_k} \to 1$ as $k \to \infty$ whence $z \in \psi_{p_0}^{-1}(D)$. We have proved: *The set of points z in S^1 such that (z^{-n_k}), $k \in \mathbb{N}$ has finitely many limit points of the form $exp(2\pi i \frac{p}{q})$, $(p, q$ integers $q > 0)$ is Lebesgue null.* We need this in **(9)** below.

(9) Let μ be a rigid measure with an associated sequence (n_k), $k \in \mathbb{N}$. Define $\mu_{(\beta)}$ by $\mu_{(\beta)}(A) = \mu(\beta A)$, $A \subseteq S^1$, where $\beta \in S^1$. Since $\hat{\mu}_{(\beta)}(n) = \beta^n \hat{\mu}(n)$, $n \in \mathbb{Z}$, we see that $\mu_{(\beta)}$ is α-rigid for every α in the limit point set of the sequence (β^{n_k}), $k = 1, 2, 3, \dots$. Now **(8)** above implies that for almost every $\beta \in S^1$ (with respect to Lebesgue measure), the limit point set of the sequence (β^{n_k}), $k = 1, 2, 3, \dots$, contains an α_0 of the form $e^{2\pi i a}$ with irrational a. Hence, in view of **(2)**, if μ is rigid then for almost every $\beta \in S^1$ (with respect to Lebesgue measure), $\mu_{(\beta)}$ is α-rigid for every α in S^1. This gives us a method of constructing α-rigid measures from rigid measures. Since every discrete probability measure is rigid, there exist rigid and α-rigid measures whose closed supports are all of S^1. Since every probability measure absolutely continuous with respect to an α-rigid measure is α-rigid, we conclude that α-rigid measures are dense in $\mathcal{P}$.

(10) The set of α-rigid measures in $\mathcal{P}$ is given by

$$\bigcap_{m=1}^{\infty} \bigcap_{l=1}^{\infty} \bigcup_{n=l}^{\infty} \left\{ \mu \in \mathcal{P} : | \hat{\mu}(n) - \alpha | < \frac{1}{m} \right\}$$

and the same holds with $\mathcal{P}$ replaced by $\mathcal{P}_C$. Combined with **(9)** above we see that α-rigid measures form a dense G_δ set in $\mathcal{P}$ and $\mathcal{P}_C$.

(11) **Exercise.** Let $D(n_1, n_2, n_3, \ldots)$ be as in **(8)**. Show that the set $\{\mu \in \mathcal{P} : \mu$ is not carried by $D(n_1, n_2, n_3, \ldots)\}$ is residual in $\mathcal{P}$.

(12) The discussion of **(8)** and **(9)** can be simplified if we use a classical result of H. Weyl ([7] p 140) which states that given a sequence $n_1 < n_2 < n_3 < \ldots$, of natural numbers, for a.e. $z \in S^1$, the sequence z^{-n_k}, $k = 1, 2, 3, \ldots$, is uniformly distributed with respect to the Lebesgue measure on S^1.

(13) A compact subset E of S^1 is called a Dirichlet set if its indicator function 1_E is a uniform limit of a sequence of continuous characters of S^1. Clearly every probability measure supported on a Dirichlet set is a rigid measure. A Borel subset of S^1 is called weak Dirichlet if every probability measure supported on it is rigid. The set $D(n_1, n_2, n_3, \ldots)$ is one example of a weak Dirichlet set but not every weak Dirichlet set is of this type. We will come across weak Dirichlet sets in the following chapters.

7.18. For a discussion on rigid measures, rigid automorphisms, and their connection with L^∞ eigenvalues of non-singular automorphisms we refer the reader to K. Schmidt [5], K. Schmidt and P. Walters [6], H. Furstenberg and B. Weiss [3], J. Aaronson [1]. The exposition of this chapter is borrowed from J. R. Choksi and M. G. Nadkarni [2].

Chapter 8

Baire Category Theorems of Ergodic Theory

Isometries of $L^p(X, \mathcal{B}, m)$

In this chapter we will describe a topology on the class of non-singular auto-morphisms on a measure space and discuss the Baire category of some naturally occurring subclasses of it. We shall make a similar study of the class of measure preserving automorphisms.

8.1. Let $(X, \mathcal{B}, m)$ be a standard probability space. Let $\mathcal{G} = \mathcal{G}(m)$ denote the group of non-singular automorphisms on $(X, \mathcal{B}, m)$ and $\mathcal{M} = \mathcal{M}(m)$ the subgroup of $\mathcal{G}$ of those automorphisms which preserve m. We identify automorphisms which differ only on a set of measure zero.

8.2. For $\tau \in \mathcal{G}$, let T_τ be defined by

$$T_\tau f = \frac{dm_\tau}{dm} f \circ \tau, \ f \in L^1(X, \mathcal{B}, m) \text{ where } m_\tau = m \circ \tau.$$

The operator T_τ is an invertible isometry of $L^1(X, \mathcal{B}, m)$.

8.3. Every invertible isometry of $L^1(X, \mathcal{B}, m)$ which preserves non-negative functions is of this form. This follows as a corollary of the following theorem of Lamperti proved in Royden's book on Real Analysis [20].

8.4 Theorem. *Let $1 \leq p < \infty$, $p \neq 2$, and let T be a linear transformation of $L^p(X, \mathcal{B}, m)$ into itself such that $\| Tf \|_p = \| f \|_p$ for all $f \in L^p(X, \mathcal{B}, m)$. Then there is a Borel function τ of X into X and an $h \in L^p(X, \mathcal{B}, m)$ such that*

$$Tf = hf \circ \tau, \quad for \ all \ f \in L^p(X, \mathcal{B}, m).$$

*The function h is uniquely determined (to within a set of measure zero) and τ
is uniquely determined to within a set of measure zero on the set where h ≠ 0.
For any Borel set E we have, with F = τ⁻¹E,*

$$\int_F | h |^p \, dm = \int_E dm. \tag{1}$$

8.5. Now, if T is an invertible isometry of $L^1(X, \mathcal{B}, m)$ which takes positive
functions to positive functions, we apply the above theorem to $L^1(X, \mathcal{B}, m)$.
Then $T1 = h \geq 0$ and (1) above tells us that $h = \frac{dm_\tau}{dm}$. Further τ is invertible
because T is invertible.

Strong Topology on Isometries

8.6. Let $\mathcal{L}(L^1, L^1)$ denote the class of all linear operators of norm ≤ 1 from
$L^1(X, \mathcal{B}, m)$ into $L^1(X, \mathcal{B}, m)$ equipped with the strong operator topology. In
this topology, $T_n \to T$ if and only if for all $f \in L^1(X, \mathcal{B}, m)$ $T_n f \to Tf$ in the
L^1 norm.

8.7. The strong operator topology on $\mathcal{L}(L^1, L^1)$ admits a complete separable
metric. To see this let $\{E_1, E_2, \ldots\}$ be a countable algebra which generates
$\mathcal{B}$. Then $T \in \mathcal{L}(L^1, L^1)$ is completely determined by the values of T at 1_{E_i},
$i \in \mathbb{N}$. The mapping $\xi : T \to (T1_{E_1}, T1_{E_2}, \ldots)$ of $\mathcal{L}$ into the countable cartesian
product $(L^1)^\mathbb{N}$ is one-one. Moreover $T_n \to T$ in the strong operator topology
if and only if for each k, $T_n 1_{E_k} \to T1_{E_k}$ (in L^1 norm), i.e., $\xi(T_n) \to \xi(T)$ in
$(L^1)^\mathbb{N}$. Thus $\mathcal{L}$ is embedded as a closed subset of $(L^1)^\mathbb{N}$, hence admits a complete
metric. One such metric is

$$d(T, S) = \sum_{n=1}^{\infty} \frac{1}{2^n} \| T1_{E_n} - S1_{E_n} \|, \ T, S \in \mathcal{L}.$$

8.8. The collection of invertible isometries in $\mathcal{L}(L^1, L^1)$ may be given a complete
separable metric by defining

$$\rho(S, T) = d(S, T) + d(S^{-1}, T^{-1}),$$

as the distance between two invertible isometries S and T.

Coarse and Uniform Topologies on $\mathcal{G}(m)$

8.9. The collection of invertible isometries in $\mathcal{L}(L^1, L^1)$ which take positive
functions to positive functions forms a complete separable metric space when
the metric ρ of **8.8** is restricted to this collection. Moreover this collection forms
a topological group under this metric. Let G denote the group of invertible
isometries in $\mathcal{L}(L^1, L^1)$ which preserve non-negative functions. Let us return

to the group $\mathcal{G}$. In view of the discussion in **8.2** and **8.3** the map $\tau \to T_\tau$ is a one-one map of $\mathcal{G}$ onto G. We give $\mathcal{G}$ the topology which makes the map $\tau \to T_\tau$ from $\mathcal{G}$ onto G a homeomorphism. This topology is metrisable in a complete separable fashion; a complete metric is given by

$$\rho(\tau, \sigma) = \rho(T_\tau, T_\sigma) = \sum_{n=1}^{\infty} (\| T_\tau 1_{E_n} - T_\sigma 1_{E_n} \| + \| T_\tau^{-1} 1_{E_n} - T_\sigma^{-1} 1_{E_n} \|).$$

8.10. The topology defined above on the group $\mathcal{G}$ is called the coarse topology. It depends on m only through the measure class of m as we show below.

8.11. Let λ be a finite measure on $\mathcal{B}$ whose null sets agree with those of m. Then clearly $\mathcal{G}(\lambda)$ and $\mathcal{G}(m)$ are the same. We can define the coarse topology on $\mathcal{G} = \mathcal{G}(\lambda)$ using the measure λ. We show that the topology remains unchanged. Let

$$Pf = \frac{d\lambda}{dm} f, \ f \in L^1(X, \mathcal{B}, \lambda).$$

Then P is an isometry from $L^1(X, \mathcal{B}, \lambda)$ onto $L^1(X, \mathcal{B}, m)$. Let $\tau \in \mathcal{G}(\lambda)$ and define

$$(S_\tau f)(x) = \frac{d\lambda_\tau}{d\lambda}(x) f(\tau x), \ f \in L^1(X, \mathcal{B}, \lambda).$$

Since

$$\frac{d\lambda \circ \tau}{dm \circ \tau} = \frac{d\lambda}{dm} \circ \tau,$$

we have $P^{-1} T_\tau P = S_\tau$. Hence $\tau_n \to \tau$ in the coarse topology of $\mathcal{G}(\lambda)$ if and only if $\tau_n \to \tau$ in the coarse topology of $\mathcal{G}(m)$.

8.12. On $\mathcal{G} = \mathcal{G}(m)$ there is another topology, finer than the coarse topology and called the uniform topology, defined by the metric

$$\delta(\sigma, \tau) = m(\{x : \sigma^{-1} x \neq \tau^{-1} x\}).$$

This metric is complete and equivalent to the metric

$$\delta_1(\sigma, \tau) = m(\{x : \sigma x \neq \tau x\}).$$

8.13. We see that the uniform topology is finer than the coarse topology through the following inequality.

$$\| T_\tau 1_B - T_\sigma 1_B \|_1 \leq 2m(B \cap \{x : \tau^{-1} x \neq \sigma^{-1} x\})$$

$$\leq 2m(\{x : \tau^{-1} x \neq \sigma^{-1} x\}).$$

To prove this inequality consider

$$\| T_\tau 1_B - T_\sigma 1_B \|_1 = \int_X \left| \frac{dm_\tau}{dm} 1_B \circ \tau - \frac{dm_\sigma}{dm} 1_B \circ \sigma \right| dm$$

$$= \int_{\tau^{-1} E} + \int_{\tau^{-1}(X - E)},$$

where E denotes the set of points $\{x : \tau^{-1}x = \sigma^{-1}x\}$. Note that

$$\tau^{-1}E = \sigma^{-1}E$$

and that $\tau x = \sigma x$ for $x \in \tau^{-1}E$. For any $A \subset \tau^{-1}E$,

$$\int_A \frac{dm_\tau}{dm}dm = m(\tau A) = m(\sigma A) = \int_A \frac{dm_\sigma}{dm}dm,$$

whence on $\tau^{-1}E$, $\frac{dm_\tau}{dm}(x) = \frac{dm_\sigma}{dm}(x)$ so that

$$\int_{\tau^{-1}E} \left| \frac{dm_\tau}{dm}1_B \circ \tau - \frac{dm_\sigma}{dm}1_B \circ \sigma \right| dm = 0. \qquad (2)$$

Now

$$\int_{\tau^{-1}(X-E)} \left| \frac{dm_\tau}{dm}1_B \circ \tau - \frac{dm_\sigma}{dm}1_B \circ \sigma \right| dm$$
$$\leq \quad m_\tau(\tau^{-1}B \cap \tau^{-1}(X-E)) + m_\sigma(\sigma^{-1}B \cap \tau^{-1}(X-E))$$
$$= \quad m(B \cap (X-E)) + m(B \cap \sigma(\tau^{-1}(X-E))).$$

Since $\sigma(\tau^{-1}E) = \tau(\tau^{-1}E) = E$ we have $\sigma(\tau^{-1}(X-E)) = X-E$. Thus

$$\| T_\tau 1_B - T_\sigma 1_B \|_1 \quad \leq \quad 2m(B \cap (X-E))$$
$$\leq \quad 2m(X-E)$$
$$\leq \quad 2m(\{x : \tau^{-1}x \neq \sigma^{-1}x\}),$$

as was required to be proved.

8.14. It is clear from **8.13** that if $\tau_n \to \tau$ in the uniform topology then $\tau_n \to \tau$ in the coarse topology. A subclass of $\mathcal{G}(m)$ dense in $\mathcal{G}(m)$ in the uniform topology is therefore dense in the coarse topology.

8.15. Recall Rokhlin's lemma. It states that if τ has no periodic points, then given $\varepsilon > 0$ and an integer $n \geq 1$, there exists a set $A \in \mathcal{B}$ such that $A, \tau A, \ldots, \tau^{n-1}A$ are pairwise disjoint and

$$m\left(X - \bigcup_{i=0}^{n-1} T^i A\right) \leq \varepsilon.$$

This lemma at once shows that periodic automorphisms are dense in $\mathcal{G}(m)$ in uniform topology, hence, also in the coarse topology.

8.16. We assume henceforth that $m\{x\} = 0$ for all $x \in X$.

8.17. Given a $\tau \in \mathcal{G}(m)$ the class $\{\phi\tau\phi^{-1} : \phi \in \mathcal{G}(m)\}$ is called the conjugacy class of τ. It is clear from Rokhlin's lemma that the conjugacy class of an aperiodic $\tau \in \mathcal{G}(m)$ is dense in the class of all aperiodic elements in $\mathcal{G}(m)$ with respect to the uniform topology, hence also dense with respect to the coarse topology. It can be shown that with respect to the coarse topology the conjugacy class of an aperiodic τ is in fact dense in all of $\mathcal{G}(m)$.

8.18. We will now discuss some other equivalent definitions of the coarse topology. Consider $L^p(X, \mathcal{B}, m), 1 \leq p < \infty$ and $\tau \in \mathcal{G}(m)$. We define

$$(T_\tau f)(x) = \left(\frac{dm_\tau}{dm}(x) \right)^{1/p} f(\tau x), \ f \in L^p(X, \mathcal{B}, m);$$

T_τ is an invertible isometry of $L^p(X, \mathcal{B}, m)$ and one can define a topology on $\mathcal{G}(m)$, which we call the p-coarse topology, by requiring that $\tau_n \to \tau$ if and only if for all $f \in L^p(X, \mathcal{B}, m)$, $T_{\tau_n} f \to T_\tau f$ in the L^p norm.

8.19. The map $\phi : \tau \to T_\tau$ is one-one from $\mathcal{G}(m)$ onto invertible isometries of $L^p(X, \mathcal{B}, m)$ which preserve positive functions. The map ϕ is a group isomorphism and the p-coarse topology makes this map a homeomorphism; $\mathcal{G}(m)$ is thus a complete separable metric group in the p-coarse topology. The map $i : \tau \to \tau$ from $\mathcal{G}$ with p-coarse topology to $\mathcal{G}$ with q-coarse topology $(p \geq q)$ is a continuous onto group isomorphism, hence a homeomorphism by a theorem of Banach and Kuratowski [1]. Thus all the p-coarse topologies $1 \leq p < \infty$ are the same.

8.20. The class $\mathcal{M}(m)$ of measure preserving automorphisms is a closed subset of $\mathcal{G}(m)$ in the coarse topology, hence also in the uniform topology.

8.21. A classical result of Halmos and Rokhlin states that the collection of weakly mixing automorphisms in $\mathcal{M}(m)$ forms a dense G_δ set and hence it is of the second category in the sense of Baire. There has been subsequent work which calculates the Baire category of various subclasses of $\mathcal{M}(m)$ and $\mathcal{G}(m)$. We will discuss a unified approach to these results and collect a number of them under one roof.

Baire Category of Classes of Unitary Operators

8.22. To this end we first digress and calculate the Baire category of various classes of unitary operators on a complex separable Hilbert space $\mathcal{H}$. Let $\mathcal{U}$ denote the collection of unitary operators on $\mathcal{H}$ equipped with the weak operator topology (which on $\mathcal{U}$ is also the strong operator topology). It is a Polish topology on $\mathcal{U}$. Fix a complete orthonormal basis $\phi_i, i \in \mathbb{N}$, in $\mathcal{H}$. For $U \in \mathcal{U}$, let F_U denote the spectral measure of U and let μ_U denote the probability measure:

$$\mu_U(A) = \sum_{n=1}^{\infty} \frac{1}{2^n} (F_U(A)\phi_n, \phi_n), \ A \subseteq S^1,$$

which is the maximal spectral type of U (μ_U depends on the complete orthonormal sequence $\phi_n, n \in \mathbb{N}$, but its measure class is independent of the choice of $\phi_n, n \in \mathbb{N}$). The map $f : U \to \mu_U$ is continuous from $\mathcal{U}$ to $\mathcal{P}$, where $\mathcal{P}$ is the collection of probability measures on the Borel subsets of S^1 equipped with the topology of weak convergence. (See Chapter 7). Since the map f is continuous, $f^{-1}(C)$ is G_δ in $\mathcal{U}$ whenever $C \subseteq \mathcal{P}$ is G_δ. Moreover, for any $C \subseteq \mathcal{P}$ which is measure class invariant (i.e., if $\mu \in C$ and if $\nu \in \mathcal{P}$ has the same null sets as μ, then $\nu \in C$), then $f^{-1}(C)$ is conjugation invariant, i.e., $U \in f^{-1}(C)$ and $W \in \mathcal{U}$, then $WUW^{-1} \in f^{-1}(C)$.

8.23 Proposition. *Let $U \in \mathcal{U}$ be such that the closed support of μ_U is S^1. Then the set $\{WUW^{-1} : W \in \mathcal{U}\}$ of conjugates of U is dense in $\mathcal{U}$. More generally if the closed support of μ_U is $K \subseteq S^1$, then the set of conjugates of U is dense in the set of $V \in \mathcal{U}$ whose maximal spectral type is supported on K.*

Proof. We prove the proposition only for the case when the closed support of μ_U is all of S^1, the general case being similar. Since operators in $\mathcal{U}$ with simple discrete spectrum are dense in $\mathcal{U}$, it is enough to prove that every V with discrete spectrum can be approximated by conjugates of U. For this it is sufficient to show that if $h_1, h_2, \ldots, h_n$ are orthonormal eigenvectors of V with eigenvalues $\lambda_1, \lambda_2, \ldots, \lambda_n$, then given $\varepsilon > 0$ there exists a $W \in \mathcal{U}$ such that $\| W^{-1}UWh_j - \lambda_j h_j \| < \varepsilon, j = 1, 2, \ldots, n$. But this follows because $\lambda_1, \ldots, \lambda_n$ are approximate eigenvalues of U (note that closed support of μ_U is all of S^1), and if $k_j, j = 1, 2, \ldots, n$, are orthonormal vectors such that $\| Uk_j - \lambda_j k_j \| < \varepsilon, j = 1, 2, \ldots, n$, then W can be chosen to map h_j to k_j, $j = 1, 2, \ldots, n$. This completes the proof.

8.24. Let us recall from chapter **7** that

(a) the atom free elements $\mathcal{P}_C$ of $\mathcal{P}$ form a dense G_δ in $\mathcal{P}$.

(b) for any $\nu \in \mathcal{P}$ the collection $\nu^\perp = \{\mu \in \mathcal{P}_C : \mu \perp \nu\}$ forms a dense G_δ in $\mathcal{P}_C$.

(c) The collection $S = \{\mu \in \mathcal{P}_C : \mu^p \perp \mu^q, p \neq q, p, q \in \mathbb{N}\}$ is dense G_δ in $\mathcal{P}_C$, where μ^p denotes the p-fold convolution of μ.

(d) For μ in $\mathcal{P}$, $\mu^{(p)}$ denotes the measure $\mu \circ \psi_p^{-1}$, where $\psi_p(z) = z^p$, $z \in S^1, p \in \mathbb{Z}$. The collection

$$R = \{\mu \in \mathcal{P}_C : \mu^{(p)} \perp \mu^{(q)}, \ p \neq q, \ p, q \in \mathbb{N}\}$$

is dense G_δ in $\mathcal{P}_C$.

(e) A $\mu \in \mathcal{P}$ is said be α-rigid ($\alpha \in S^1$) if there is an increasing sequence $n_k, k = 1, 2, 3 \ldots$ of natural numbers such that $\hat{\mu}(n_k) \to \alpha$ as $k \to \infty$. A 1-rigid measure is simply called rigid. The class of atom free α-rigid and 1-rigid measures form a dense G_δ set in $\mathcal{P}_C$. We denote α-rigid measures in $\mathcal{P}_C$ by T^α and the class of rigid measures in $\mathcal{P}_C$ by T.

8.25 Theorem. *The sets*

(a) $f^{-1}(\mathcal{P}_C) = \{U \in \mathcal{U} : \mu_U \in \mathcal{P}_C\} =_{def} \mathcal{U}_C$

(b) $f^{-1}(\nu^{\perp}) = \{U \in \mathcal{U}_C : \mu_U \perp \nu\} =_{def} \mathcal{U}_{\nu^{\perp}}$

(c) $f^{-1}(R) = \{U \in \mathcal{U}_C : \mu_U^{(p)} \perp \mu_U^{(q)}, p \neq q, p, q \in \mathbb{N}\} =_{def} R_1$

(d) $f^{-1}(S) = \{U \in \mathcal{U}_C : \mu_U^p \perp \mu_U^q, p \neq q, p, q \in \mathbb{N}\} =_{def} S_1$

(e) $f^{-1}(T^{\alpha}) = \{U \in \mathcal{U}_C : \mu_U \;\; is \;\; \alpha - rigid\} =_{def} T_1^{\alpha}$

(f) $f^{-1}(T) = \{U \in \mathcal{U}_C : \mu_U \;\; is \;\; rigid\} =_{def} T_1$

are conjugation invariant dense G_{δ} in $\mathcal{U}_C$. The corresponding sets in $\mathcal{U}$ are residual.

Proof. The class $\mathcal{P}_C$ is dense G_{δ} in $\mathcal{P}$ and measure class invariant. Hence the class $f^{-1}(\mathcal{P}_C)$ is G_{δ} and conjugation invariant. It is dense since it contains a U with closed support of μ_U all of S^1. The classes $\nu^{\perp}, R, S, T^{\alpha}, T$ are all dense G_{δ} in $\mathcal{P}_C$ and measure class invariant, hence their inverse images under f are G_{δ} in $\mathcal{U}_C$ and conjugation invariant. Each of these classes contains a U with closed support of μ_U all of S^1. As before these classes are dense. The last statement is obvious. It is to be noted with regard to (c) that the measure classes $[\mu_{U^p}]$ and $[\mu^{(p)}]$ are the same. The theorem is proved.

8.26 Theorem. *The set $\{U \in \mathcal{U} : U \;\; has \;\; multiplicity \;\; one\}$ is dense G_{δ} in $\mathcal{U}$.*

Proof. Let $\mathcal{A}$ be a countable set of elements in $\mathcal{H}$, and let $\mathcal{C}$ be another set in $\mathcal{H}$, not necessarily countable. Let C be a set of complex numbers. Let $W \subseteq \mathcal{U}$ be defined as follows: $U \in W \Leftrightarrow$ given $f_1, f_2, \ldots, f_n \in \mathcal{A}, \varepsilon > 0, \exists g \in \mathcal{C}, N \in \mathbb{N}$, and $\lambda_{i,-N}, \ldots, \lambda_{i,N} \in C, i = 1, 2, \ldots, n$ such that

$$\left\| f_i - \sum_{j=-N}^{N} \lambda_{i,j} U^j g \right\| < \varepsilon, i = 1, 2, \ldots, n.$$

We see that W is a G_{δ} set as follows: Let $\mathcal{A} = \{f_1, f_2, \ldots\}$. Given $m, n, N \in \mathbb{N}$ and $g \in \mathcal{C}$, let $A(n, N, m, g)$ be the set of $U \in \mathcal{U}$ such that

$$\exists \lambda_{i,k} \in C, 1 \leq i \leq n, -N \leq k \leq N, \left\| f_i - \sum_{k=-N}^{N} \lambda_{i,k} U^k g \right\| < \frac{1}{m}.$$

The set $A(n, N, m, g)$ is open in $\mathcal{U}$ and we have

$$W = \bigcap_{m=1}^{\infty} \bigcap_{n=1}^{\infty} \bigcup_{N=1}^{\infty} \bigcup_{g \in \mathcal{C}} A(n, N, m, g).$$

Hence W is a G_{δ} set in $\mathcal{U}$. Now let $\mathcal{A}$ be any countable dense set in $\mathcal{H}$. Set $\mathcal{C} = \mathcal{H}$ and $C = \mathbb{C}$. The class W thus obtained, which we now denote by W_1, is G_{δ}. Since $\mathcal{A}$ is dense we see that $U \in W_1$ if and only if given $f_1, \ldots, f_n \in \mathcal{H}$ and $\varepsilon > 0$, there exists a g in $\mathcal{H}$ such that for each i,

$$\inf_h \| f_i - h \| < \varepsilon,$$

infimum being taken as h runs over the closed linear span of $U^k g, k \in \mathbb{Z}$. The following lemma of Katok and Stepin , which is a special case of theorem **4.3**, now permits us to conclude that $\mathcal{W}_1$ consists of precisely those $U \in \mathcal{U}$ which have spectral multiplicity one. Since the set of $U \in \mathcal{U}$ with simple discrete spectrum is dense in $\mathcal{U}$, the theorem follows.

8.27 Lemma (Katok-Stepin). *$U \in \mathcal{U}$ has spectral multiplicity greater than one if and only if there exist $\phi_1, \phi_2 \in \mathcal{H}$ such that for all $g \in \mathcal{H}$,*

$$d^2(\phi_1, H(g)) + d^2(\phi_2, H(g)) \geq 1,$$

where $H(g)$ is the subspace spanned $U^k g, k \in \mathbb{Z}$. The vectors ϕ_1 and ϕ_2 can be chosen to be orthogonal.

8.28 Corollary. The set $\{U \in \mathcal{U} : U$ has spectral multiplicity one$\}$ is dense G_δ in $\mathcal{U}_C$ and $\mathcal{U}$.

Baire Category of Classes of Non-Singular Automorphisms

8.29. Let us now return to $\mathcal{G}(m)$ and associate with each $\tau \in \mathcal{G}(m)$ the operator U_τ on $L^2(X, \mathcal{B}, m)$:

$$(U_\tau f)(x) = \left(\frac{dm_\tau}{dm}(x) \right)^{1/2} f(\tau x), f \in L^2(X, \mathcal{B}, m).$$

We equip $\mathcal{G}(m)$ with the coarse topology. Then the map $\xi : \tau \to U_\tau$ is a continuous map from $\mathcal{G}(m)$ into $\mathcal{U}$. (Recall that all the p-coarse topologies for $1 \leq p < \infty$ coincide, hence the coarse topology is the same as the topology which makes the map ξ continuous with $p = 2$).

8.30 Theorem. *All the sets*

(a) $\xi^{-1}(\mathcal{U}_C) = \{\tau \in \mathcal{G} : \mu_{U_\tau} \in \mathcal{P}_C\} =_{def} \mathcal{G}_C(m) = \mathcal{G}_C,$
(b) $\xi^{-1}(\mathcal{U}_\nu^\perp) = \{\tau \in \mathcal{G}_C : \mu_{U_\tau} \perp \nu\},$
(c) $\xi^{-1}(R_1) = \{\tau \in \mathcal{G}_C : \mu_{U_\tau}^{(p)} \perp \mu_{U_\tau}^{(q)}, \ p \neq q, p, q \in \mathbb{N}\},$
(d) $\xi^{-1}(S_1) = \{\tau \in \mathcal{G}_C : \mu_{U_\tau}^p \perp \mu_{U_\tau}^q, p \neq q, p, q \in \mathbb{N}\},$
(e) $\xi^{-1}(T_1) = \{\tau \in \mathcal{G}_C : \mu_{U_\tau}$ *is rigid*$\}$
(f) $\xi^{-1}(\mathcal{W}_1) = \{\tau \in \mathcal{G}_C : U_\tau$ *has spectral multiplicity one*$\},$
(g) $\xi^{-1}(T_1^\alpha) = \{\tau \in \mathcal{G}_C : \mu_{U_\tau}$ *is* $\alpha - rigid\},$

are G_δ and conjugation invariant. Further the sets (a)-(f) are dense while the set (g) is empty for $\alpha \neq 1, -1$.

Proof. All the sets **(a)** - **(g)** are G_δ being the inverse images of G_δ sets under a continuous map. All the classes **(a)** - **(g)** are clearly conjugation invariant in $\mathcal{G}(m)$. Now any member of $\mathcal{G}_C$ is aperiodic hence its conjugacy class is dense in $\mathcal{G}_C$ (indeed in $\mathcal{G}$) by our discussion in **8.17**. Further all the classes **(a)**-**(f)** are non-empty, hence dense in $\mathcal{G}_C$ and $\mathcal{G}$. If μ_{U_τ} is α-rigid for $\alpha \neq 1, -1$ then its measure class is not symmetric (see chapter **3**) whereas the measure class of μ is always symmetric, hence the class **(g)** is empty for $\alpha \neq 1, -1$.

8.31. All the members of the class **(f)** are ergodic so that ergodic members of $\mathcal{G}(m)$ are residual in $\mathcal{G}(m)$. This result is due to J. R. Choksi and S. Kakutani [3] who in fact prove that the class of ergodic τ in $\mathcal{G}(m)$ is dense G_δ in $\mathcal{G}(m)$.

8.32 Exercise. Prove the non-emptiness of the classes **(a)-(f)**.

Baire Category of Classes of Measure Preserving Automorphisms

8.33. We now consider $\mathcal{M}(m)$, the class of measure preserving automorphisms on (X, B, m). Let $L_0^2(X, \mathcal{B}, m)$ be the collection of functions in $L^2(X, \mathcal{B}, m)$ with zero integral. Every $f \in L^2(X, \mathcal{B}, m)$ can then be written as $f = (f - \int_X f) + \int_X f = g + c$, where $\int_X g = 0$ and c is a constant.

Now $\tau_n \to \tau$ in the coarse topology on $\mathcal{M}$ if and only if $f \circ \tau_n \to f \circ \tau$ for all $f \in L^2(X, \mathcal{B}, m)$. Since $f \circ \tau_n = g \circ \tau_n + c$ we see that $\tau_n \to \tau$ if and only if $g \circ \tau_n \to g \circ \tau$ for all $g \in L_0^2(X, \mathcal{B}, m)$. We see therefore that the map $\phi : \tau \to U_\tau$, where U_τ is defined on $L_0^2(X, \mathcal{B}, m)$ by $U_\tau f = f \circ \tau$, is a continuous one-one map from $\mathcal{M}(m)$ into the group $\mathcal{U}$ of unitary operators on $\mathcal{H} = L_0^2(X, \mathcal{B}, m)$.

8.34 Theorem. *The sets*

(a) $\phi^{-1}(\mathcal{U}_C) = \{\tau \in \mathcal{M} : \mu_{U_\tau} \in \mathcal{P}_C\} =_{def} \mathcal{M}_C$,
(b) $\phi^{-1}(\mathcal{U}_{\nu^\perp}) = \{\tau \in \mathcal{M}_C : \mu_{U_\tau} \perp \nu\} = \mathcal{M}_{\nu^\perp}$,
(c) $\phi^{-1}(R_1) = \{\tau \in \mathcal{M}_C : \mu_{U_\tau^p} \perp \mu_{U_\tau^q}, \ p \neq q, \ p, \ q \in \mathbb{N}\}$,
(d) $\phi^{-1}(S_1) = \{\tau \in \mathcal{M}_C : \mu_{U_\tau}^p \perp \mu_{U_\tau}^q, \ p \neq q, \ p, \ q \in \mathbb{N}\}$,
(e) $\phi^{-1}(W_1) = \{\tau \in \mathcal{M}_C : U_\tau \ \text{has spectral multiplicity one}\}$,
(f) $\phi^{-1}(T_1) = \{\tau \in \mathcal{M}_C : \mu_{U_\tau} \ \text{is rigid}\}$,

are all dense G_δ in $\mathcal{M}_C$ and in $\mathcal{M}$.

Proof. All the classes are G_δ being the inverse images of G_δ sets under a continuous map. They are also conjugation invariant and non-empty. Any member of $\mathcal{M}$ is aperiodic and its conjugacy class is dense in $\mathcal{M}_C$ (and in $\mathcal{M}$) in the coarse topology. The theorem follows.

8.35 Exercise. Prove that all the classes in the above theorem are non-empty.

8.36. The class $\mathcal{M}_C$ is precisely the set of measure preserving τ which are weakly mixing. This class was shown to be G_δ in $\mathcal{M}$ by Halmos [10]. Now mixing automorphisms fall in the complement of the set $\phi^{-1}(T_1)$, hence form a meager set in $\mathcal{M}_C$ and $\mathcal{M}$. (Rokhlin [19]). This proves the existence of weakly mixing automorphisms which are not mixing. For a concrete example of a weakly mixing automorphism which is not mixing we have to either appeal to the theory of Gauss automorphisms as discussed in Cornfeld, Sinai and Fomin [5], or to Chacon's automorphism (See K. Petersen [17]). The class (e) was shown to be dense G_δ by A. Katok and A. Stepin [14] and (f) is shown to be dense G_δ in A. Katok [15]. Further A. M. Stepin [21] shows that some of these properties can be realised within the class of smooth systems preserving a smooth measure.

8.37 Corollary (Rokhlin). *Given a $\nu \in \mathcal{P}$, the set $\{\tau \in \mathcal{M}_C : \mu_{U_\tau} \ll \nu\}$ is meager in $\mathcal{M}_C$.*

Proof. The class in question falls in the complement of the set $\mathcal{M}_{\nu^\perp}$ which is dense G_δ.

Baire Category and Joinings

8.38. We now relate parts (b) and (c) of theorem **8.34** to certain Baire category results of A. del Junco [6] concerning the class of automorphisms disjoint in the sense of Furstenberg.

8.39 Definition (Furstenberg [7]). Two measure preserving automorphisms σ and τ on $(X, \mathcal{B}, m)$ are said to be disjoint if whenever there is a third measure preserving automorphism ω and two ω-invariant sub σ-algebras $\mathcal{A}$ and $\mathcal{C}$ of $\mathcal{B}$ such that $\omega \mid_{\mathcal{A}}$ is isomorphic to σ and $\omega \mid_{\mathcal{C}}$ is isomorphic to τ, then $\mathcal{A}$ and $\mathcal{C}$ are independent.

8.40 Theorem (Hahn and Parry[9]). *If two measure preserving automorphisms σ and τ are such that U_σ and U_τ on $L_0^2(X, \mathcal{B}, m)$ have mutually singular maximal spectral type, then σ and τ are disjoint.*

From this theorem we have the inclusions:

(1) Fix $\sigma \in \mathcal{M}$, then

$$\{\tau \in \mathcal{M} : \tau \text{ disjoint from } \sigma\} \supseteq \{\tau \in \mathcal{M}_C : \mu_{U_\tau} \perp \mu_{U_\sigma}\}.$$

(2) $\{\tau \in \mathcal{M} : \tau^p \text{ disjoint from } \tau^q, \mid p \mid \neq \mid q \mid, \ p, \ q \in \mathbb{Z}\} \supseteq \{\tau \in \mathcal{M}_C : \mu_{U_\tau^p} \perp \mu_{U_\tau^q}, \mid p \mid \neq \mid q \mid, \ p, \ q \in \mathbb{Z}\}.$

From theorem **8.34** (b),(c) and the above inclusions we conclude:

8.41 Theorem (A. del Junco). *The set of measure preserving automorphisms disjoint from a fixed measure preserving automorphism is residual in $\mathcal{M}$ in the coarse topology. The set of measure preserving automorphisms all whose powers are disjoint is residual in $\mathcal{M}$ in the coarse topology.*

In del Junco [6] these sets are in fact shown to be dense G_δ.

8.42. For more on uniform topology see N. Friedman [8]. For **8.26** see J. Hawkins and A. Robinson [13]. The exposition in this chapter has relied mainly on (i) J. R. Choksi and S. Kakutani [3], (ii) J. R. Choksi and M. G. Nadkarni [4]. We refer to B. Simon [21] where similar results are proved concerning the spectrum of self-adjoint operators.

8.43 Asides. In his paper entitled " Probability and Physical Systems" [2] (p 369), G. D. Birkhoff defines metrical transitivity (which is same as ergodicity) and says: "The importance of this idea arises from the fact that, almost certainly, recurrent physical systems are in general metrically transitive, although this is very difficult to prove". Again at the end of the same paper (p 379) "The outstanding problem concerning physical systems from the point view of probability is that of determining to what extent recurrent systems are transitive. It is probable that in general there is metrical transitivity. It would be a distinct advance even to establish that there is metrical transitivity in the case of the geodesics on closed surfaces of negative curvature."

The conjecture of Birkhoff that in general there is metric transitivity was given a precise meaning by J. C. Oxtoby and S. Ulam who proved in their 1941 paper [16] that in the group of measure preserving homeomorphisms on $[0,1]^n, n \geq 2$, equipped with the topology of uniform convergence, the set of ergodic measure preserving homeomorphisms is dense G_δ. Soon after, Halmos [11] discussed the class of all measure preserving automorphisms on a finite measure space and proved the residuality of the class of weakly mixing automorphisms under a suitable topology. Similar results were also proved by Rokhlin [19]. The calculation of Baire category of suitable subclasses of measure preserving or non-singular automorphism has remained a topic of research since the publication of the paper of Oxtoby and Ulam. The ergodicity of geodesic flows on a surface of constant negative curvature was proved by E. Hopf, thus confirming the second of Birkhoff's conjectures mentioned above. (See Cornfeld, Sinai, and Fomin [5]). The ergodic theory of manifolds of negative curvature has remained a topic of deep ergodic theory ever since.

Chapter 9

Translations of Measures on the Circle

A Theorem of Weil and Mackey

Given a measure μ on S^1 we can associate with it two sets:

$$A(\mu) = \{t \in S^1 : \mu_t \text{ and } \mu \text{ are not mutually singular}\},$$

$$H(\mu) = \{t \in S^1 : \mu_t \text{ and } \mu \text{ have the same null sets}\}.$$

Here μ_t denotes the measure μ translated by t: $\mu_t(A) = \mu(tA)$, for any Borel subset A of S^1.

These sets occur naturally in the spectral theory of non-singular transformations. The aim of this chapter is to expose some general facts about these sets and apply them to prove a result of B. Host (Theorem **8.26.**) on marginal measures. This result is needed in the next chapter for the solution of the multiple mixing problem when the spectrum is singular. First we recall a theorem of A. Weil as improved by G. W. Mackey on measure and topology in groups (**9.1–9.4**).

9.1. Let G be a group with a Borel structure $\mathcal{B}$ such that $(G, \mathcal{B})$ is a standard Borel space and the map $(g, h) \to gh^{-1}$ is measurable. If $A \in \mathcal{B}$ and $h \in G$, then this map is one-one on the set $A \times \{h\}$ with image $A \cdot h^{-1}$ which is therefore a set in $\mathcal{B}$. Applying a similar argument to $\{e\} \times A$ we see that A^{-1} is Borel whenever A is Borel.

9.2. Let m be a non-trivial σ-finite measure on $\mathcal{B}$ such that for all $g \in G$ and $A \in \mathcal{B}$, $m(gA) = m(A)$. Then, according to a theorem of A. Weil, there is a unique locally compact topology $\mathcal{T}$ on G such that $(G, \mathcal{T})$ is a topological group, $\mathcal{T}$ generates $\mathcal{B}$, and m is a Haar measure on G.

9.3. It is not necessary to assume that m is invariant under G in the result quoted above. As observed by G. W. Mackey, if m is σ-finite and quasi-invariant under G then there exists a σ-finite μ in the measure class of m which is invariant under G. We see this as follows:

(a) Quasi-invariance of m means that for all $g \in G$, the measure m_g and m have the same null sets where $m_g(A) = m(gA), A \in \mathcal{B}$.

Let $\mathcal{D}_n = \{D_1^n, D_2^n, \ldots, D_{2^n}^n\}, n = 1, 2, 3, \ldots$, be a sequence of partitions, $\mathcal{D}_{n+1}$ refining $\mathcal{D}_n$, such that $\bigcup_{n=1}^{\infty} \mathcal{D}_n$ generates $\mathcal{B}$. Define

$$\phi_n(g, x) = \begin{cases} \frac{m_g(D_k^n)}{m(D_k^n)} & \text{if } x \in D_k^n, \ m(D_k^n) \neq 0), \\ 0 & \text{otherwise.} \end{cases}$$

The ϕ_n's are jointly measurable and for each g

$$\phi_n(g, x) \to \frac{dm_g}{dm}(x), \quad a.e. \ m$$

by the martingale convergence theorem. If we define

$$\phi(g, x) = \lim_{n \to \infty} \phi_n(g, x)$$

whenever the limit exists, $\phi(g, x) = 0$ otherwise, then ϕ is jointly measurable in g and x and for all $g \in G$,

$$\phi(g, x) = \frac{dm_g}{dm}(x) \quad a.e. \ m.$$

(b) For all $g, h \in G$,

$$\phi(gh, x) = \phi(h, x)\phi(g, hx) \quad a.e. \ m.$$

To see this let $A \in \mathcal{B}$, and consider

$$
\begin{aligned}
\int_A \phi(h, x)\phi(g, hx)dm &= \int_A \phi(g, hx)\frac{dm_h}{dm}(x)dm(x) = \int_A \phi(g, hx)dm_h(x) \\
&= \int_{hA} \phi(g, y)dm(y) = \int_{hA} \frac{m_g}{dm}(y)dm(y) \\
&= \int_{hA} dm_g(y) = m(ghA) \\
&= \int_A \frac{dm_{gh}}{dm}(x)dm(x) = \int_A \phi(gh, x)dm.
\end{aligned}
$$

Since this holds for all $A \in \mathcal{B}$, we see that

$$\phi(gh, x) = \phi(h, x) \times \phi(g, hx) \quad a.e. \ m. \tag{1}$$

(c) If we assume that (1) holds for all $x \in G$ rather than almost every where, then setting $x = e$ we get $\phi(g, h) = \phi(gh, e)(\phi(h, e))^{-1}$.
Write $\xi(h) = \phi(h, e)$ then $\phi(g, h) = \xi(gh)(\xi(h))^{-1}$, for all $g, h \in G$. Define $\mu(A) = \int_A \frac{1}{\xi(x)} dm$. Then μ is a σ-finite measure on G and μ and m have the same null sets since ϕ is a positive finite valued function. Further

$$\mu(gA) = \int_{gA} \frac{1}{\xi(x)} dm(x) = \int_A \frac{1}{\xi(gx)} dm_g(x) =$$

$$\int_A \frac{1}{\xi(gx)} \frac{dm_g}{dm}(x) dm(x) = \int_A \frac{1}{\xi(gx)} \frac{\xi(gx)}{\xi(x)} dm =$$

$$\int_A \frac{1}{\xi(x)} dm(x) = \mu(A),$$

which shows that μ is invariant under G.

(d) We now show, as an application of Fubini's theorem, that we can find a jointly measurable version of ϕ satisfying (1) for all g, h and x. The map $(g, h, x) \to \phi(gh, x)$ is jointly measurable in g, h, x and satisfies for all g, h the cocycle identity (1) m a.e. in x. Hence by Fubini's theorem, for a.e. $g, h, x \in G \times G \times G$ (with respect to the measure $m \times m \times m$),

$$\phi(gh, x) = \phi(h, x)\phi(g, hx).$$

Again by Fubini's theorem there exists an $x' \in G$ such that

$$\phi(gh, x') = \phi(h, x')\phi(g, hx') \quad a.e. \quad g, h.$$

We set $hx' = y$ and $\xi(y) = \phi(yx'^{-1}, x')$. Then

$$\phi(g, y) = \frac{\phi(gh, x')}{\phi(h, x')} = \frac{\phi(gyx'^{-1}, x')}{\phi(yx'^{-1}, x')} = \frac{\xi(gy)}{\xi(y)} \quad a.e. \quad (g, y).$$

Thus the cocycle ϕ is a coboundary. The function ξ is finite positive valued hence also the function $\frac{1}{\xi}$. We summarise the above considerations below.

9.4 Theorem (Weil-Mackey). *Let G be a group with a standard Borel structure $\mathcal{B}$ such that the map $(g, h) \to gh^{-1}$ is measurable. Let m be a non-trivial σ-finite measure on $\mathcal{B}$ such that for all $g \in G$ the measures m and m_g have the same null sets. Then there is a unique topology $\mathcal{T}$ on G such that $(G, \mathcal{T})$ is a locally compact second countable topological group whose open sets generate $\mathcal{B}$ and whose Haar measure has the same class of null sets as m.*

For an exposition of Weil's theorem we refer the reader to [1], [8], and to [5] for Mackey's improvement of it.

The Sets $A(\mu)$ and $H(\mu)$ and Their Topologies

9.5. We now come to the main theme of this chapter, namely, measures on S^1 in relation to the translations on the circle. Let μ be a finite measure on Borel subsets $\mathcal{B}$ of S^1. We introduce the sets

$$A(\mu) = \{t \in S^1 : \mu_t \text{ and } \mu \text{ are not mutually singular}\},$$

$$H(\mu) = \{t \in S^1 : \mu_t \text{ and } \mu \text{ have the same null sets}\}.$$

Clearly $H(\mu) \subseteq A(\mu)$.

9.6. Let μ_t^a denote the part of μ_t absolutely continuous with respect to μ and let

$$f(t, x) = \frac{d\mu_t^a}{d\mu}(x).$$

By adapting the method of **9.3(a)** it is possible to choose, for each t, a version of $\frac{d\mu_t^a}{d\mu}$ such that f is jointly measurable in t and x. Now $A(\mu) = \{t : \int_{S^1} f(t, x)d\mu > 0\}$ and since $\int_{S^1} f(t, s)d\mu$ is measurable in t we see that $A(\mu)$ is a Borel subset of S^1. Similarly $H(\mu)$ is a Borel subset of S^1.

9.7. We have

$$\| \mu_t - \mu \| = \sup \left\{ \left| \int_{S^1} (f(xt) - f(x))d\mu \right| : f \text{ continuous}, | \text{ f } | \le 1 \right\}.$$

As a function of t, $\| \mu_t - \mu \|$ is lower semicontinuous so that

$$A(\mu) = \{t : \| \mu_t - \mu \| < 2 \| \mu \|\}$$

is an F_σ set. We will see later that $H(\mu)$ is also an F_σ set.

9.8. We topologise $A(\mu)$ and $H(\mu)$ as follows: Identify $L^1(S^1, \mu)$ with the class all complex measures λ on $\mathcal{B}$ absolutely continuous with respect to μ. For $t \in S^1$ let λ_t denote the translate of λ by t and define $L_t \lambda = \lambda_t^a = $ part of λ_t absolutely continuous with respect to μ. The operator L_t is linear and bounded on $L^1(S^1, \mu)$, $\| L_t \| \le 1$. The class of all bounded linear operators T on $L^1(S^1, \mu)$ of norm ≤ 1 is separable under the strong topology in which a sequence $T_n \to T$ if and only if for all $f \in L^1(S^1, \mu)$, $T_n f$ converges in norm to Tf. We topologise $A(\mu)$ by requiring that a sequence (t_n) in $A(\mu)$ converge to t in $A(\mu)$ if $L_{t_n} \to L_t$ in the strong topology. The topology on $H(\mu)$ is defined similarly. The topology on $A(\mu)$ is separable metrisable and that on $H(\mu)$ admits, in addition, a complete metric.

9.9 Proposition. *If $t_n \in A(\mu)$ converges to $t \in A(\mu)$ in the above topology then $t_n \to t$ in the usual topology. Further*

$$\frac{d\mu_{t_n}^a}{d\mu} \to \frac{d\mu_t^a}{d\mu}$$

in $L^1(S^1, \mu)$.

Proof. $t_n \to t$ in the above described topology means that $L_{t_n} \to L_t$ in the strong operator topology. Hence $L_{t_n}\mu \to L_t\mu$ in $L^1(S^1, \mu)$, i.e.,

$$\frac{d\mu_{t_n}^a}{d\mu} \to \frac{d\mu_t^a}{d\mu} \quad \text{in } L^1(S^1, \mu).$$

To prove the first assertion let $t_n \in A(\mu)$ converge to $t \in A(\mu)$ in the topology described above. Then for *any* continuous function f, whose translate by t we denote by f_t, we have

$$f_{t_n}\frac{d\mu_{t_n}}{d\mu} \to f_t\frac{d\mu_t}{d\mu}$$

in $L^1(S^1, \mu)$ from which we conclude that $t_n \to t$ in the usual topology.

9.10 Proposition. *Let $t_n, t \in A(\mu)$ and assume that $t_n \to t$ in the usual topology of S^1 and*

$$\frac{d\mu_{t_n}^a}{d\mu} \to \frac{d\mu_t^a}{d\mu} \quad \text{in } L^1(S^1, \mu).$$

Then $t_n \to t$ in the topology of $A(\mu)$.

Proof. It is enough to show that if $\lambda = fd\mu$ where f is a continuous function on S^1, then

$$\frac{d\lambda_{t_n}^a}{d\mu} \to \frac{d\lambda_t^a}{d\mu} \quad \text{in } L^1(S^1, \mu) \text{ as n} \to \infty.$$

The assumptions of the proposition clearly implies this.

9.11. Note that the injection $t \to t$ of $A(\mu)$ (with the new topology) into S^1 (with the usual topology) is continuous.

Groups Generated by Dense Subsets of $A(\mu)$; Their Properties

9.12 Theorem. *Let μ be a finite measure on $(S^1, \mathcal{B})$. Then there exists a countable subgroup D of S^1 such that for every $\lambda \in L^1(S^1, \mu)$ with $\lambda_t \in L^1(S^1, \mu)$:*

(i) *$\lambda_t(A) = \lambda(A)$ for every D-invariant Borel set A,*
(ii) *$\lambda(A - t^{-1}A) = 0$ for every D-invariant Borel set A.*

Proof. Let J be a countable dense set in $A(\mu)$ (in the new topology of $A(\mu)$). Let D be the subgroup of S^1 generated by J. For $d \in D$, and any D invariant Borel set A

$$(L_d\lambda)(A) = \lambda_d^a(A) \le \lambda_d(A) = \lambda(A),$$

since A is D-invariant. Since D is dense in $A(\mu)$, for any $t \in A(\mu)$ and any D-invariant A

$$(L_t \lambda)(A) = \lim_{d_n \to t} (L_{d_n} \lambda)(A) \leq \lambda(A),$$

where $d_n \in D$, $d_n \to t$ in $A(\mu)$.

If $\lambda_t \in L(S^1, \mu)$, as is part of the hypothesis, then $\lambda_t(A) \leq \lambda(A)$ for every D-invariant Borel set A. Similarly $\lambda_t(S^1 - A) \leq \lambda(S^1 - A)$, i.e., $\lambda_t(A) \geq \lambda(A)$, whence $\lambda_t(A) = \lambda(A)$. To prove (ii) consider $\lambda' = 1_A d\lambda$. As A is D-invariant, for every $d \in D$, $\lambda'_d(S^1 - A) = 0$, hence also $(L_d \lambda')(S^1 - A) = 0$. By denseness of D, $(L_t \lambda')(S^1 - A) = 0$. Since $\lambda'_t \ll \lambda_t \ll \mu$, we have $L_t \lambda' = \lambda'_t$ and $\lambda'_t(S^1 - A) = 0$. Therefore λ'_t is concentrated on A; equivalently λ' is concentrated on At^{-1};

$$\lambda(A - At^{-1}) = 0.$$

9.13. If D is as in the above theorem (with respect to μ) then for all $t \in S^1$ and for all D-invariant Borel sets A, $(1_A d\mu)_t$ and $1_{A^c} d\mu$ are mutually singular. Indeed if $t \in D$ then the conclusion is valid because A is D-invariant. If $t \in A(\mu)$ but not in D then the conclusion is valid because D is dense in $A(\mu)$ in the topology of $A(\mu)$. Finally for t not in $A(\mu)$, μ and μ_t are mutually singular by definition of $A(\mu)$.

9.14. The subgroup D satisfying the properties of **9.12.** is not unique. However the σ-algebra generated by the D-invariant Borel sets together with the μ null sets is independent of the choice of D and depends only on μ. To see this let D and D' be two subgroups with the properties of proposition **9.12.** Let A' be a D' invariant Borel set and consider $A = \bigcup_{\alpha \in D} \alpha A'$. Then A is D-invariant and also D'-invariant. We show that $\mu(A - A') = 0$. To see this note that by **9.13.**, for all $t \in S^1$,

$$(1_{A'} d\mu)_t \perp (1_{A-A'}) d\mu. \tag{1}$$

If $\mu(A - A') > 0$, then by definition of A, for some $\alpha \in D$,

$$\mu((A - A') \cap \alpha A') > 0,$$

which contradicts (1) above. This proves the assertion.

Ergodic Measures on the Circle Group

9.15 Definition. Let μ be a finite measure on $(S^1, \mathcal{B})$ and D a countable subgroup of S^1. We say that μ is D-ergodic if for every D-invariant Borel set A, either $\mu(A) = 0$ or $\mu(S^1 - A) = 0$. (Note that we are not assuming that μ is invariant or quasi-invariant with respect to the D-action).

9.16. Any measure absolutely continuous with respect to Haar measure is ergodic with respect to any countable dense subgroup of S^1. A discrete measure is ergodic with respect to the group generated by the point masses. There are also continuous measures singular to Haar measure which are ergodic with respect to suitable subgroups of S^1. As an example we show that the Cantor measure on the usual Cantor ternary set is ergodic with respect to the group generated by the end points of the deleted intervals. Let I denote the interval $0 \le x < 1$. Expand x in its ternary expansion $x = \sum_{i=1}^{\infty} \frac{1}{3^i} x_i$, $x_i = 0, 1$ or 2. We make the expansion unique by requiring that the number of terms in the expansion be minimum. Let $E = \{x : x_i = 0 \text{ or } 2\}$, which is the usual Cantor ternary set but for a small modification which is done for convenience rather than necessity. Write for $x \in E$,

$$\psi(x) = \sum_{i=1}^{\infty} \frac{1}{2^i} x_i;$$

ψ is a one-one continuous function from E onto I. Further ψ is strictly increasing on E. Define a measure μ on E by $\mu(A) = L(\psi^{-1}(A))$, where L is the Lebesgue measure on I; μ is the usual Cantor measure on E. Let G be the subgroup of real numbers generated by members in E having finitely many terms in their ternary expansions. We show that μ is G-ergodic. Let A be a Borel subset of $\mathbb{R}$ which is G-invariant. Let $E_1 = E \cap A$ and consider $\psi(E_1)$. We show that $\psi(E_1)$ is invariant (mod 1) under the group H consisting of all real numbers having finitely many terms in their binary expansion. (As before we make the expansion unique by requiring that the number of terms in the expansion be minimum). Let $\alpha \in \psi(E_1)$; $\alpha = 0.\alpha_1\alpha_2\alpha_3 \cdots$ be its binary expansion. Let $h \in H$ have binary expansion $h = 0.h_1h_2h_3 \cdots h_n$. Then it is clear that $\alpha + h$ (mod 1) and α agree in their binary expansion from term $n + 1$ onwards. Hence, from the way ψ is defined, $\psi^{-1}(\alpha + h) - \psi^{-1}(\alpha) = g$ (say) has finitely many terms in its ternary expansion so that $g \in G$. Now $\alpha \in \psi(E_1)$, therefore $\psi^{-1}(\alpha) \in E_1$. Consequently $\psi^{-1}(\alpha+h) = \psi^{-1}(\alpha)+g \in E_1$. (Recall that $E_1 = E \cap A$, and A is invariant under G). We have thus $\alpha + h = \psi(\psi^{-1}(\alpha + h)) \in \psi(E_1)$. This shows that $\psi(E_1)$ is invariant under translations by members of H (mod 1). Since the Lebesgue measure on $[0, 1)$ is ergodic with respect any countable dense subgroup of $\mathbb{R}$, we see that $L(\psi(E_1)) = 0$ or 1. Consequently $\mu(E_1) = 0$ or 1, which proves the ergodicity of μ under G.

9.17 Exercise. Let E be an uncountable Borel subset of $\mathbb{R}$ which is independent over the field of rational numbers. Show that a continuous probability measure supported on E cannot be ergodic with respect to any countable subgroup of $\mathbb{R}$.

9.18 Proposition. *For any finite measure μ on $(S^1, \mathcal{B})$, the following are equivalent:*

(i) *μ is D-ergodic for any countable subgroup D of S^1 such that $D \cap A(\mu)$ is dense in $A(\mu)$.*

(ii) *μ is D-ergodic for some countable subgroup D of S^1.*

(iii) *For all $A, B \in \mathcal{B}$ with $\mu(A), \mu(B) > 0$ there exists a t such that $\mu(tA \cap B) >$*
    0.

Proof. Clearly (i) $\Rightarrow$ (ii) $\Rightarrow$ (iii). We complete the proof by showing that (iii)
$\Rightarrow$ (i). Suppose (iii) holds and let D be a countable subgroup generated by a
dense subset of $A(\mu)$. If A is a D-invariant set of positive measure we show that
$\mu(S^1 - A) = 0$. We know that for all $t \in S^1$, $(1_A d\mu)_t$ and $1_{A^c} d\mu$ are mutually
singular. If $\mu(S^1 - A) \neq 0$, then by (iii) for some t, $\mu(tA \cap (S^1 - A)) > 0$, a
contradiction. Thus μ is D-ergodic.

9.19. It is not known whether $D \cap A(\mu)$ is dense in $A(\mu)$ in the topology of
$A(\mu)$ whenever μ is D-ergodic.

9.20 Exercise. Show that if μ and ν are D-ergodic then $\mu * \nu$ is D-ergodic.

9.21 Exercise. Let μ be ergodic with respect to a countable group D generated
by a subset of $A(\mu)$. Let $D = \{d_1, d_2, d_3, \ldots\}$ and let $\nu = \sum_{k=1}^{\infty} \frac{1}{2^k} \mu_{d_k}$. Then ν
is D-ergodic and $H(\nu)$ is the group generated by $A(\mu)$.

9.22. We have discussed the ergodicity of μ without requiring μ to be invariant
or quasi-invariant under the group action. It is customary to discuss ergodicity
under a non-singular group action. If μ is quasi-invariant and ergodic under
the action of countable subgroup $D \subseteq S^1$ then for all $t \in S^1$, μ_t is also quasi-
invariant and ergodic under D. Since any two quasi-invariant ergodic measures
are either mutually singular or mutually absolutely continuous, the measures
μ and μ_t are either mutually singular or mutually absolutely continuous. For a
quasi-invariant measure μ, therefore, $H(\mu) = A(\mu)$.

9.23 Theorem. *Let μ be a probability measure on $(S^1, \mathcal{B})$, quasi-invariant and
ergodic with respect to a countable subgroup D of S^1. Then $\mu(H(\mu)) > 0$ if
and only if μ has the same null sets as either the Haar measure on S^1 or the
cardinality measure on D.*

Proof. First we prove the $\Rightarrow$ part. Suppose $\mu(H(\mu)) > 0$. Consider $G = H(\mu)$
with its Borel structure $\mathcal{B}_G = G \cap \mathcal{B}$ and the measure μ. (Note that since
$\mu(H(\mu)) > 0$, by ergodicity, μ is supported on $H(\mu)$). $(G, \mathcal{B}_G)$ is a standard
Borel space, the map $(g, h) \to gh^{-1}$ is measurable and μ on $\mathcal{B}_G$ is quasi-
invariant under the G-action. By the Weil-Mackey theorem **9.4**, there exists a
unique second countable locally compact topology $\mathcal{T}$ on G under which $(G, \mathcal{T})$
is a topological group, $\mathcal{B}_G$ is the σ-algebra generated by $\mathcal{T}$, and μ and the
Haar measure on G have the same null sets. The injection map $i : G \to S^1$,
$i(x) = x$, is a Borel measurable group isomorphism, hence a continuous group
isomorphism. Now from the structure theory of locally compact Abelian groups
we know that there exists an open subgroup of $(G, \mathcal{T})$ isomorphic to $\mathbb{R}^n \times K$,
for some n and some compact group K. We note that $n = 0$, for otherwise
the injection map i provides a one-one continuous isomorphism of $\mathbb{R}^n$ into S^1

which is impossible. Now consider K. If K is finite, then since K is open in G, $\mathcal{T}$ must be the discrete topology on $G = H(\mu)$. Since $(G, \mathcal{T})$ is separable, G must be countable. Further G must be equal to D, for if $G \neq D$, then D and $G - D$ are disjoint D-invariant sets each of positive measure, contradicting the ergodicity of μ. Suppose now that K is infinite. Then K is dense in S^1. Further K is a measurable subgroup of S^1. The injection map i is, as before, continuous so that $i(K)$ is a compact dense subgroup of S^1, i.e., $K = S^1$. Since μ and the Haar measure on K have the same null sets the $\Rightarrow$ part is proved. The implication $\Leftarrow$ is obvious.

9.24 Exercise. Let μ be a probability measure on $(S^1, \mathcal{B})$, quasi-invariant and ergodic with respect to a countable dense subgroup of S^1. For each Borel support E of μ write $H_E = \{t : tE \text{ is a support of } \mu\}$. Show that H_E is a Borel set and that $H(\mu)$ is the intersection of the H_E's taken over all Borel supports E. Also show that $\mu(H_E)$ is either 0 or $\mu(S^1)$.

9.25. Suppose μ is continuous, singular with respect to Haar measure on S^1, and D-ergodic for some countable subgroup D. A theorem of G. Brown and W. Moran states that then μ and $\mu * \mu$ are mutually singular. It is possible to deduce theorem **9.23.** as a corollary of this theorem. For a relatively simple proof of the theorem of Brown and Moran see B. Host and F. Parreau. [3].

9.26 Corollary. *If ν is a probability measure on $(S^1, \mathcal{B})$, ergodic with respect to a countable subgroup D of S^1 (but not necessarily quasi-invariant under the D-action) and if $\nu(A(\nu)) > 0$, then ν is either discrete or absolutely continuous with respect to the Haar measure on S^1.*

Proof. The measure $\mu = \sum_{n=1}^{\infty} \frac{1}{2^n} \nu_{d_n}$, where $d_1, d_2, d_3, \ldots$ is an enumeration of D, is quasi-invariant and ergodic under the D-action. Further $\mu(H(\mu)) > \nu(A(\nu)) > 0$. By theorem **9.23** μ is either discrete or absolutely continuous with respect to Haar measure on S^1. Since $\nu \ll \mu$, the same holds for ν.

A Theorem on Marginal Measures

9.27. As an application of the above considerations **9.12, 9.23** we now prove a theorem on marginal measures which is needed in the solution of the multiple mixing problem in the singular spectrum case. We need some notation. Let

  (i) S denote the symmetry map $s \to s^{-1}$ of S^1,
  (ii) $S \times S$ denotes the symmetry map $(s, t) \to (s^{-1}, t^{-1})$ of $S^1 \times S^1$,
  (iii) U denotes the map which permutes co-ordinates: $(s, t) \to (t, s)$,
  (iv) V denotes the map $(s, t) \to (s, t^{-1}s^{-1})$.

All these maps are their own inverses. Further $S \times S$ commutes with U and V. We also let π_1 and π_2 denote the projections $\pi_1(s, t) = s$, $\pi_2(s, t) = t$.

9.28 Theorem (B. Host). *Let λ be a probability measure on the Borel subsets of $S^1 \times S^1$ such that*

(a) λ, $\lambda \circ (S \times S)$, $\lambda \circ U$, $\lambda \circ V$ *are all the same,*
(b) λ *is absolutely continuous with respect to the product of its projections:*
$$\lambda \ll \lambda \circ \pi_1^{-1} \times \lambda \circ \pi_2^{-1}.$$

Then each projection of λ is the sum of a discrete measure and a measure absolutely continuous with respect to the Haar measure on S^1.

Proof. Since λ is invariant under permutation of co-ordinates the projections $\lambda \circ \pi_1^{-1}$, $\lambda \circ \pi_2^{-1}$ of λ are the same, say equal to μ. Note that μ is S invariant. We have to show that μ is the sum of a discrete measure and a measure absolutely continuous with respect to Haar measure on S^1. Let $F(\cdot, \cdot)$ denote the Radon-Nikodym derivative of λ with respect $\mu \times \mu$. For every $s \in S^1$, let $\lambda(s, \cdot)$ denote the measure

$$\lambda(s, A) = \int_A F(s, t) d\mu(t), \ A \in \mathcal{B}.$$

For μ a.e. s the measure $\lambda(s, \cdot)$ is finite and non-null. For every bounded Borel function f on $S^1 \times S^1$,

$$\int_{S^1 \times S^1} f d\lambda = \int_{S^1} \left(\int_{S^1} f(s, t) \lambda(s, dt) \right) \mu(ds).$$

We now claim that for μ a.e. s

$$\lambda(s, \cdot) \circ R_s^{-1} = \lambda(s, \cdot) \circ S, \tag{1}$$

where R_s is the map $t \to st$ on S^1. Since $\lambda \circ V = \lambda$, for every pair of continuous functions f, g on S^1:

$$\int_{S^1} \left(\int_{S^1} f(t) \lambda(s, dt) \right) g(s) \mu(ds)$$
$$= \int_{S^1} \int_{S^1} g(s) f(t) \lambda(dsdt)$$
$$= \int_{S^1} \int_{S^1} g(s) f(t) \lambda \circ V(dsdt)$$
$$= \int_{S^1} \int_{S^1} g(s) f(s^{-1} t^{-1}) \lambda(dsdt)$$
$$= \int_{S^1} \left(\int_{S^1} f(s^{-1} t^{-1}) \lambda(s, dt) \right) g(s) \mu(ds)$$
$$= \int_{S^1} \left(\int_{S^1} f(t) \lambda \circ R_s^{-1} S(s, dt) \right) g(s) \mu(ds).$$

Comparing the first and the last expressions in the above chain of equalities and recalling that the equality holds for all continuous f on S^1, we see that (1) holds for μ a.e. s.

Let D be a subgroup of S^1 satisfying the properties of theorem **9.12** with respect to μ: For each D-invariant Borel set A, for each $\nu \ll \mu$ and for all t with $\nu \circ R_t^{-1} \ll \mu$, we have $\nu(A - R_t A) = 0$. Now for μ a.e. s, $\lambda(s, \cdot) \ll \mu$ and by (1) $\lambda(s, \cdot) \circ R_s^{-1} = \lambda(s, \cdot) \circ S \ll \mu \circ S = \mu$. Hence by the definition of D: for μ a.e. s and for each $A \in \mathcal{D}$,

$$\lambda(s, A - R_s A) = 0,$$

and since μ is symmetric,

$$\lambda(s, A - R_s^{-1}A) = 0, \tag{2}$$

where $\mathcal{D}$ denotes the σ-algebra of D-invariant Borel sets.

Let P denote the map $(s, t) \to (st)$ and let A, B be two D-invariant Borel sets with $A \cap B = \emptyset$. Then

$$\lambda(A \times B - P^{-1}(B)) = \int_A \lambda(s, B - R_s^{-1}B)\mu(ds) = 0$$

by (2).

Similarly $\lambda(A \times B - P^{-1}(A)) = 0$. Since $(P^{-1}A) \cap (P^{-1}B) = \emptyset$ we conclude that $\lambda(A \times B) = 0$. Thus

$$\forall A, B \in \mathcal{D} \text{ with } A \cap B = \emptyset, \ \lambda(A \times B) = 0. \tag{3}$$

We now claim that the restriction of μ to $\mathcal{D}$ is purely atomic. If not there is a set $A \in \mathcal{D}$ with $\mu(A) > 0$ such that for all $\varepsilon > 0$ there exists a finite partition $\{A_i : i \in I\}$ of A into D-invariant Borel sets with $\mu(A_i) < \varepsilon$ for each i. Now from (3) $\lambda(A \times S^1) = \lambda(A \times A)$ and $\lambda(A \times A)$ is positive since $\mu(A)$ is positive. Again by (3) $\lambda \mid_{A \times A}$ is supported on $K = \bigcup A_i \times A_i$ with $\mu \times \mu(K) = \sum(\mu(A_i))^2 \leq \varepsilon$. Since ε is arbitrary we see that $\lambda \mid_{A \times A}$ is supported on a set with $\mu \times \mu$ measure zero. Since $\lambda \ll \mu \times \mu$, a contradiction is reached and the restriction of μ to $\mathcal{D}$ is proved to be purely atomic.

Let E be one such atom and let $\nu = \mu \mid_E$. Then ν is D-ergodic. We show that $\nu(A(\nu)) > 0$. By (3), $\lambda(E \times (S^1 - E)) = 0$. Therefore for ν a.e. s, $\lambda(s, \cdot)$ is concentrated on E, hence absolutely continuous with respect to ν. Further for ν a.e. s, $\lambda(s, \cdot)$ is concentrated on $R_s^{-1}E$ by (2). Therefore $\lambda(s, \cdot) \circ R_s^{-1}$ is concentrated on E. As $\lambda(s, \cdot) \circ R_s^{-1} \ll \mu$ by (1), we conclude that $\lambda(s, \cdot) \circ R_s^{-1} \ll \nu$ for ν a.e. s. Finally, since $\lambda(s, \cdot)$ is non-null for ν a.e. s, ν and $\nu \circ R_s^{-1}$ are not mutually singular, i.e., $\nu(A(\nu)) > 0$. Since ν is D-ergodic we conclude by **9.23** that ν is either discrete or absolutely continuous with respect to the Haar measure on S^1. The conclusion concerning μ follows since the restriction of μ to each atom of $\mathcal{D}$ is either discrete or absolutely continuous with respect to the Haar measure on S^1.

9.29. The set $A(\mu)$ is introduced in B. Host, J.-F. Méla, F. Parreau [3] and its properties as discussed here are borrowed from B. Host, J.-F. Méla, F. Parreau [3] and B. Host [2]. Example **9.16** and Theorem **9.23** are borrowed from V. Mandrekar and M. Nadkarni[6].

Chapter 10

B. Host's Theorem

Pairwise Independent and Independent Joinings of Automorphisms

10.1. Let τ_1 and τ_2 be two measure preserving automorphisms on standard probability spaces $(X_1, \mathcal{B}_1, m_1)$ and $(X_2, \mathcal{B}_2, m_2)$. A measure m on $\mathcal{B}_1 \times \mathcal{B}_2$ is said to be a joining of τ_1 and τ_2 if m is invariant under $\tau_1 \times \tau_2 : (x_1, x_2) \to (\tau_1 x_1, \tau_2 x_2)$, and the projection of m on X_1 is m_1 and on X_2 is m_2. The product measure $m_1 \times m_2$ is a joining of τ_1 and τ_2 but the systems may admit other joinings. The measure $m_1 \times m_2$ is called the independent joining of τ_1 and τ_2.

10.2. A joining of a measure preserving automorphism with itself is called a self joining. If τ is defined on a standard probability space $(X, \mathcal{B}, \mu)$ and preserves μ then, for any $n \in \mathbb{Z}$, the measure m_n defined on $\mathcal{B} \times \mathcal{B}$ by

$$m_n(A \times B) = \mu(A \cap \tau^{-n} B),$$

is a self joining. The measure m_n is, in fact, the measure μ transferred to the graph of τ^n by the map $x \to (x, \tau^n x)$. Further

$$m_n(\tau A \times \tau B) = \mu(\tau A \cap \tau^{-n+1} B) = m_n(A \times B),$$

$$m_n(X \times B) = \mu(B), \quad m_n(A \times X) = \mu(A),$$

so that m_n is indeed a self joining of τ.

10.3. Continuing with the discussion of **10.2**, suppose that τ is a mixing automorphism. Then, as $n \to \infty$, the self joining m_n of **10.2** converges to the independent joining of τ with itself:

$$m_n(A \times B) = \mu(A \cap \tau^{-n} B) \to \mu(A) \cdot \mu(B).$$

Whether τ is mixing or not, $\mu(A \cap \tau^{-n} B)$, $n \in \mathbb{N}$, always has a subsequence which converges. Let $\mathcal{F}$ be a countable subalgebra of $\mathcal{B}$ which generates $\mathcal{B}$ and let $n_k, k \in \mathbb{N}$, be a fixed increasing sequence of natural numbers such that for all $A, B \in \mathcal{F}$, $\mu(A \cap \tau^{-n}B)$ converges over the subsequence $n_k, k \in \mathbb{N}$. If we define

$$m(A \times B) = \lim_{k \to \infty} \mu(A \cap \tau^{-n_k}B), \quad A, B \in \mathcal{F},$$

then m is a self joining of τ.

10.4. Let $\tau_1, \tau_2, \ldots, \tau_r$ be r measure preserving automorphisms on standard probability spaces $(X_1, \mathcal{B}_1, m_1), \ldots, (X_r, \mathcal{B}_r, m_r)$. A joining m of $\tau_1, \ldots, \tau_r$ is a measure on $\mathcal{B}_1 \times \cdots \times \mathcal{B}_r$ invariant under

$$\tau_1 \times \cdots \times \tau_r : (x_1, \ldots, x_r) \to (\tau_1 x_1, \ldots, \tau_r x_r)$$

and whose projections on the co-ordinate spaces $X_1, \ldots, X_r$ are the measures $m_1, \ldots, m_r$ respectively. The measure $m_1 \times m_2 \times \cdots \times m_r$ is a joining of $\tau_1, \ldots, \tau_r$ which is called the independent joining of $\tau_1, \ldots, \tau_r$.

10.5. If $\tau_1 = \tau_2 = \cdots = \tau_r = \tau$ (so that all the τ_i's are defined on the same probability space), then a joining of the τ_i's is called an r-fold self joining of τ or an r-fold joining of τ with itself.

10.6 Definition. A joining m of $\tau_1, \tau_2, \ldots, \tau_r$ is said to be pairwise independent if the projections of m on $X_i \times X_j$ for every pair i, j $(i \neq j)$ is an independent joining of τ_i and τ_j.

B. Host's Theorem: The Statement

10.7. A theorem of B. Host [2] states that if at least $r-2$ of $\tau_1, \ldots, \tau_r$ are weakly mixing with their spectra singular to Lebesgue measure, then every pairwise independent joining of $\tau_1, \ldots, \tau_r$ is the independent joining of $\tau_1, \ldots, \tau_r$. We will prove this theorem in this chapter, but first we show how this theorem permits us to prove that if τ is mixing and U_τ has spectrum singular to Lebesgue measure then τ is mixing of all orders.

Mixing Implies Multiple Mixing if the Spectrum is Singular

10.8. Let τ be a mixing automorphism on $(X, \mathcal{B}, \mu)$ with spectrum singular to Lebesgue measure. We wish to show that τ is 3-fold mixing. If it is not 3-fold mixing then there exist three sets A, B, C in $\mathcal{B}$ and two increasing sequences of positive integers $m_i, n_i, i \in \mathbb{N}$, such that $\mid m_i - n_i \mid \to \infty$ as $i \to \infty$ and $\mu(A \cap \tau^{m_i}B \cap \tau^{n_i}C)$ converges to a real number which is not $\mu(A) \cdot \mu(B) \cdot \mu(C)$. Let $\mathcal{F}$ be a countable subalgebra of $\mathcal{B}$ containing A, B, C and which generates $\mathcal{B}$. We can choose a subsequence of $(m_i, n_i), i \in \mathbb{N}$, also denoted by $(m_i, n_i), i \in \mathbb{N}$, such that for all D, E, F in $\mathcal{F}$ the quantity $\mu(D \cap \tau^{m_i} E \cap \tau^{n_i} F)$ converges to a limit as $i \to \infty$. We define m by:

$$m(D \times E \times F) = \lim_{i \to \infty} \mu(D \cap \tau^{m_i} E \cap \tau^{n_i} F), \quad D, E, F \in \mathcal{F}.$$

Then m is a three fold joining of τ which is pairwise independent (since τ is mixing) but not independent because $m(A \times B \times C) \neq \mu(A)\mu(B)\mu(C)$. This contradicts Host's theorem since τ is mixing (hence weak mixing) with spectrum singular to Lebesgue measure. Similarly we can show that if τ is mixing and has spectrum singular to Lebesgue measure then τ is r-fold mixing for every r.

B. Host's Theorem: The Proof

10.9. Suppose τ_1, τ_2, τ_3 are measure preserving automorphisms with τ_1 weakly mixing and having spectrum singular to Lebesgue measure. Suppose we have proved that every pairwise independent joining of such a system is the independent joining, then we can prove Host's theorem for any $r > 3$ as follows. Let r be greater than 3 and suppose that for every $s \leq r$ any pairwise independent joining of s measure preserving automorphisms (with at least $s - 2$ weakly mixing and having spectrum singular to Lebesgue measure) is the independent joining. Suppose now that we have $r + 1$ measure preserving automorphisms $\tau_1, \tau_2, \ldots, \tau_{r+1}$, of which $(r - 1)$ are weakly mixing with spectrum singular to Lebesgue measure. We may assume that $\tau_2, \ldots, \tau_r$ are weakly mixing with spectrum singular to Lebesgue measure. Let m be a joining of $\tau_1, \tau_2, \ldots, \tau_{r+1}$ which is pairwise independent. Then the projection of m on $X_2 \times X_3 \times \cdots \times X_r$ is a pairwise independent joining of $\tau_2, \ldots, \tau_r$, hence an independent joining by assumption. Let $Z = X_2 \times \cdots \times X_r$ and T on Z be $\tau_2 \times \cdots \times \tau_r$. The induction assumption also tells us that the projection of m on $X_1 \times X_2 \times \cdots \times X_r$ is an independent joining of $\tau_1, \tau_2, \ldots, \tau_r$. Similarly the projection of m on $X_2 \times \cdots \times X_{r+1}$ is an independent joining of $\tau_2, \tau_3, \ldots, \tau_{r+1}$. Now m is a pairwise independent joining of τ_1, T, and τ_{r+1} with T weakly mixing and spectrum singular to Lebesgue measure. Hence by Host's theorem for $r = 3$ we see that m is the independent joining of τ, T, and τ_{r+1}; hence of $\tau_1, \ldots, \tau_{r+1}$. It is therefore enough to prove Host's theorem for $r = 3$. We proceed in that direction.

10.10. Let m be a pairwise independent joining of τ_1, τ_2 and τ_3 such that for all bounded measurable functions f_1, f_2, f_3 defined on X_1, X_2, X_3 respectively,

$$\int_Y f_1(x_1)f_2(x_2)f_3(x_3)dm = \int_{X_1} f_1 dm_1 \int_{X_2} f_2 dm_2 \int_{X_3} f_3 dm_3,$$

where Y stands for cartesian product of X_1, X_2, X_3. Clearly, m is then the independent joining of $\tau_1, \tau_2,$ and τ_3. Replacing f_i by $f_i - \int_{X_i} f_i dm_i$, $i = 1, 2, 3$, we can say, equivalently, that a pairwise independent joining m is the independent joining if whenever f_1, f_2, f_3 are bounded measurable functions on X_1, X_2, X_3 respectively with vanishing integrals, the integral of $f_1 \cdot f_2 \cdot f_3$ with respect to

m vanishes. We will verify this for a pairwise independent joining m of τ_1, τ_2, τ_3 at least one of which is weakly mixing and has spectrum singular to Lebesgue measure. We will reduce this problem to a problem in harmonic analysis by showing that there exists a finite complex measure ρ on $S^1 \times S^1 \times S^1$ whose Fourier transform $\hat{\rho}$ is given by

$$\hat{\rho}(k,m,n) = \int_Y f_1(\tau_1^k x_1) f_2(\tau_2^m x_2) f_3(\tau_3^n x_3) dm, \quad k,\ m,\ n\ \in \mathbb{Z},$$

where ρ further has the properties:

(i) it is concentrated on the subgroup $H = \{(r,s,t) : rst = 1\}$,
(ii) each of the natural projections of ρ on $S^1 \times S^1$ is absolutely continuous with respect to a product measure.

Under our hypothesis on the τ_i's we will show that such a ρ is necessarily zero. (Setting $k = m = n = 0$ we see that integral of $f_1 \cdot f_2 \cdot f_3$ with respect to m will then be zero).

10.11. Recall that $m_1 \times m_2$ is invariant under the $\mathbb{Z} \times \mathbb{Z}$ action (τ_1^n, τ_2^m), $m, n \in \mathbb{Z}$. Consider the group of unitary operators $U_{n,m}$, defined by

$$(U_{n,m} f)(x_1, x_2) = f(\tau_1^n x_1, \tau_2^m x_2), \quad f \in L^2(X_1 \times X_2, m_1 \times m_2),$$

$(m,n) \in \mathbb{Z} \times \mathbb{Z}$. By the spectral theorem there exists, for each pair ϕ, ψ of functions in $L^2(X_1 \times X_2, m_1 \times m_2)$, a complex measure $\sigma_{\phi,\psi}$ on $S^1 \times S^1$ such that for all $n, m \in \mathbb{Z}$,

$$\hat{\sigma}_{\phi,\psi}(n,m) = \int \int_{X_1 \times X_2} \phi(\tau_1^n x_1, \tau_2^m x_2) \overline{\psi}(x_1, x_2) dm_1(x_1) dm_2(x_2).$$

If $\phi = \psi$ then we write σ_ϕ for $\sigma_{\phi,\phi}$. Note that σ_ϕ is non-negative and $\sigma_{\phi,\psi}$ is absolutely continuous with respect to σ_ϕ. If $\phi(x_1, x_2) = f(x_1) g(x_2)$ then $\sigma_\phi = \sigma_f \times \sigma_g$ where σ_f and σ_g are measures on S^1 whose Fourier transforms are

$$\hat{\sigma}_f(n) = \int_{X_1} f(\tau_1^n x_1) \overline{f}(x_1) dm_1, \quad \hat{\sigma}_g(n) = \int_{X_2} g(\tau_2^n x_2) \overline{g}(x_2) dm_2, n \in \mathbb{Z}.$$

(There is a slight abuse of notation here which will cause no confusion.)

10.12 Theorem. *Let m be a pairwise independent joining of τ_1, τ_2, τ_3 and let f_1, f_2, f_3 be bounded measurable functions on X_1, X_2, X_3 respectively. Then there exists a finite complex measure ρ on $S^1 \times S^1 \times S^1$ such that*

(i) $\forall\ m, n, p \in \mathbb{Z}$,

$$\hat{\rho}(m,n,p) = \int_Y f_1(\tau_1^m x_1) f_2(\tau_2^n x_2) f_3(\tau_3^p x_3) dm(x_1, x_2, x_3),$$

(ii) ρ is concentrated on the closed subgroup $H = \{(s,t,u) : stu = 1\}$,

(iii) the images of $|\rho|$ by the natural projections $\pi_{12}, \pi_{23}, \pi_{31}$ of $S^1 \times S^1 \times S^1$ onto $S^1 \times S^1$ are absolutely continuous with respect to $\sigma_{f_1} \times \sigma_{f_2}, \sigma_{f_2} \times \sigma_{f_3}, \sigma_{f_3} \times \sigma_{f_1}$ respectively.

Proof. Let $G(x_1, x_2)$ be the conditional expectation of $f_3(x_3)$ given x_1, x_2: G is a bounded measurable function characterised by (up to $m_1 \times m_2$ null sets),

$$\int_{X_1 \times X_2} G(x_1, x_2) F(x_1, x_2) d(m_1 \times m_2)(x_1, x_2)$$

$$= \int_Y f_3(x_3) F(x_1, x_2) dm(x_1, x_2, x_3)$$

for all bounded measurable functions F on $X_1 \times X_2$. Let λ denote the correlation measure $\sigma_{K, \overline{G}}$ where $K(x_1, x_2) = f_1(x_1) \cdot f_2(x_2)$. Let ρ be the image of λ under the map $(s,t) \to (s, t, s^{-1}t^{-1})$. The measure ρ is concentrated on the subgroup $H = \{(s,t,u) : stu = 1\}$. Further, for all $m, n, p \in \mathbb{Z}$,

$$\int_Y f_1(\tau_1^m x_1) f_2(\tau_2^n x_2) f_3(\tau_3^p x_3) dm = \int_Y f_1(\tau_1^{m-p} x_1) f_2(\tau_2^{n-p} x_2) f_3(x_3) dm$$

$$= \int_{X_1 \times X_2} f_1(\tau_1^{m-p} x_1) f_2(\tau_2^{n-p} x_2) G(x_1, x_2) dm_1(x_1) dm_2(x_2)$$

$$= \hat{\lambda}(m - p, n - p) = \hat{\rho}(m, n, p).$$

This proves (i) and (ii). To prove (iii) note that λ is absolutely continuous with respect to $\sigma_K = \sigma_{f_1} \times \sigma_{f_2}$. Further, it is the image by the projection π_{12} of the measure ρ which is concentrated on H. As the restriction of π_{12} to H is one-one, the measure $|\rho| \circ \pi_{12}^{-1}$ is equal to $|\lambda|$ and thus absolutely continuous with respect to $\sigma_{f_1} \times \sigma_{f_2}$. Similarly

$$|\rho| \circ \pi_{23}^{-1} \ll \sigma_{f_2} \times \sigma_{f_3},$$

$$|\rho| \circ \pi_{31}^{-1} \ll \sigma_{f_3} \times \sigma_{f_1}.$$

This proves the theorem.

We now state the key harmonic analysis result whose proof depends on theorem **9.28**.

10.13 Proposition. Let ρ be a measure on $S^1 \times S^1 \times S^1$ concentrated on $H = \{(s,t,u) : stu = 1\}$ and suppose that each of its natural projections on $S^1 \times S^1$ is absolutely continuous with respect to some product measure. Then each projection of ρ on S^1 is the sum of a discrete measure and a measure absolutely continuous with respect to Lebesgue measure on S^1.

Proof. Let S denote the symmetry map $s \to s^{-1}$ from S^1 to S^1, and let V denote the map $(s,t) \to (s, s^{-1}t^{-1})$. Each of the measures arising from ρ

by co-ordinate permutation satisfies the same properties as ρ. In addition, the average of these measures satisfies these properties since a sum of product measures is absolutely continuous with respect to some product measure. If the conclusion of the proposition holds for this average measure, it holds for ρ. We can thus restrict to the case when ρ is invariant under co-ordinate permutations. Similarly we can assume that ρ is invariant under the symmetry map $S \times S \times S$ of $S^1 \times S^1 \times S^1$. Let λ and μ denote the measures

$$\lambda = \rho \circ \pi_{12}^{-1}, \text{ where } \pi_{12}(s, t, u) = (s, t),$$

$$\mu = \rho \circ \pi_1^{-1}, \text{ where } \pi_1(s, t, u) = s.$$

The measure μ is invariant under S. We have to prove that it is the sum of a discrete measure and a measure absolutely continuous with respect to Lebesgue measure on S^1. Note that $\mu = \rho \circ \pi_2^{-1} = \rho \circ \pi_3^{-1}$ because we have assumed that ρ is invariant under permutation of co-ordinates.

The projection π_{12}, restricted to H, is one-one and its inverse is $(s, t) \rightarrow (s, t, s^{-1}t^{-1})$ which we call ξ. As ρ is concentrated on H, it is the image of λ by this mapping, i.e., $\rho = \lambda \circ \xi^{-1}$. Since ρ is invariant under permutation of co-ordinates, ρ is also the image of λ by the mapping $(s, t) \rightarrow (s, s^{-1}t^{-1}, t)$, so that $\lambda = \lambda \circ V^{-1}$. The measure λ is absolutely continuous with respect to some product measure, hence it is absolutely continuous with respect to the product of its projections. Thus the measure λ on $S^1 \times S^1$ has the properties

(i) $\lambda \circ S \times S = \lambda \circ V = \lambda$, and λ is invariant under co-ordinate permutation,
(ii) $\lambda \ll \mu \times \mu$.

By theorem **9.28** we conclude that μ is the sum of a discrete measure and a measure absolutely continuous with respect to Lebesgue measure. The theorem stands proved.

10.14 Theorem.   *Let m be a pairwise independent joining of three measure preserving automorphisms τ_1, τ_2, τ_3, at least one of which is weakly mixing with spectrum singular to Lebesgue measure. Then m is the independent joining.*

Proof.   Suppose that τ_1 is weakly mixing with spectrum singular to Lebesgue measure. Let f_1, f_2, f_3 be any three bounded measurable functions on X_1, X_2, X_3 respectively with zero expectations. Let ρ be the complex measure on $S^1 \times S^1 \times S^1$ provided by theorem **10.12**. Then $|\rho|$ satisfies the hypothesis of proposition **10.13**, hence each natural projection of $|\rho|$ on S^1 is the sum of a discrete measure and a measure absolutely continuous with respect to Lebesgue measure. On the other hand $|\rho| \circ \pi_1^{-1}$ is absolutely continuous with respect to σ_{f_1}, a continuous measure which is singular to Lebesgue measure. This shows that $\rho = 0$ and

$$\int_Y f_1(x_1) f_2(x_2) f_3(x_3) dm = \hat{\rho}(0, 0, 0) = 0.$$

Since this holds for all bounded Borel functions with vanishing integrals, and since m is a pairwise independent joining, we see that m is the independent joining.

Summing up:

10.15 Theorem. *Every pairwise independent joining of r measure preserving automorphisms is an independent joining whenever at least $r - 2$ of these automorphisms are weakly mixing with spectrum singular to Lebesgue measure.*

10.16 Corollary. *If a measure preserving automorphism on a probability space is mixing with spectrum singular to Lebesgue measure then it is mixing of all orders.*

10.17. A theorem S. Kalikow [3] says that mixing rank one automorphisms are mixing of all orders. It is not known, however, if mixing rank one automorphisms can have a Lebesgue component in their spectrum.

An Improvement and an Application

10.18. In [1] I. Assani has improved the results of this chapter and applied them to a question of H. Furstenberg on the almost every where convergence of multiple weak mixing averages.

A probability preserving system $(X, \mathcal{B}, \mu, T)$ is said to be mixing of order 3 if, for every $A_i \in \mathcal{B}, (1 \leq i \leq 3)$,

$$\mu(A_1 \cap T^{n_2} A_2 \cap T^{n_3} A_3) \to \prod_{i=1}^{3} \mu(A_i) \text{ as } n_2 \to \infty, n_3 - n_2 \to \infty.$$

A function $f_1 \in L^2(\mu)$ generates the mixing of order three property for the system $(X, \mathcal{B}, \mu, T)$ if for all functions $f_2, f_3 \in L^\infty(\mu)$,

$$\int_X f_1(x) f_2(T^{n_2} x) \cdot f_3(T^{n_3} x) d\mu \to \prod_{i=1}^{3} \int_X f_i d\mu.$$

In [1] Assani proves:

Theorem. *Let $(X_1, \mathcal{B}_1, \mu_1, T_1)$ be a weakly mixing system. Let w be a pairwise independent joining of this system with two ergodic systems $(X_2, \mathcal{B}_2, \mu_2, T_2)$ and $(X_3, \mathcal{B}_3, \mu_3, T_3)$, one of then being weakly mixing. Take $f \in L^2(\mu_1)$ and denote by Pf the projection of f onto the vector space of those functions whose spectral measure is absolutely continuous with respect to Lebesgue measure m. Then for* all $f_2, f_3 \in L^\infty(\mu)$ we have,

$$\int f(x_1)f_2(x_2)f_3(x_3)dw(x_1, x_2, x_3) = \left(\int f d\mu_1\right)\left(\int f_2 d\mu_2\right)\left(\int f_3(x_3)d\mu_3\right)$$

$$+ \int Pf(x_1)f_2(x_2)f_3(x_3)dw(x_1, x_2, x_3).$$

The following corollary and the theorem are then immediate, by taking $Pf = 0$.

Corollary. Let $(X_1, \mathcal{B}_1, \mu_1, T_1)$ be a weakly mixing system and let $f_1 \in L^2(\mu_1)$ be such that $\sigma_{f_1} \perp m$ (where σ_{f_1} is the spectral measure of f_1 and m is the Lebesgue measure). Then for all pairwise independent joinings w of $(X_1, \mathcal{B}_1, \mu_1, T_1)$ with two ergodic systems $(X_2, \mathcal{B}_2, \mu_2, T_2)$ and $(X_3, \mathcal{B}_3, \mu_3, T_3)$, one of them being weakly mixing, we have

$$\int f_1(x_1)f_2(x_2)f_3(x_3)dw = \left(\int f_1 d\mu_1\right)\left(\int f_2 d\mu_2\right)\left(\int f_3(x_3)d\mu_3\right).$$

Theorem. Let $(X, \mathcal{B}, \mu, T)$ be a mixing system and let $f_1 \in L^2(\mu)$ be such that $\sigma_{f_1} \perp m$. Then f generates the mixing of order 3 property.

These results are then used, together with other arguments, to prove the following multiple weak mixing theorem (almost every where version) which is a contribution to a question raised Furstenberg (see [1]).

Theorem. Let $(X, \mathcal{B}, \mu, T)$ be a weakly mixing dynamical system such that the restriction of T to its Pinsker algebra (largest invariant subalgebra on which the entropy of T is zero) has spectrum singular to Lebesgue measure. Then for all positive integers H, for all $f_i \in L^\infty(\mu), 1 \leq i \leq H$, the averages

$$\frac{1}{N}\sum_{n=1}^{N} f_1(T^n x)f_2(T^{2n}x)\cdots f_H(T^{Hn}x) \quad \text{converge a.e. to} \quad \prod_{i=1}^{H}\int_X f_i d\mu.$$

Chapter 11

L^∞ Eigenvalues of Non-Singular Automorphisms

The Group of Eigenvalues and Its Polish Topology

11.1   Let $(X, \mathcal{B}, m)$ be a standard probability space and let T be an ergodic non-singular automorphism on $(X, \mathcal{B}, m)$. A complex number λ is said to an L^∞ eigenvalue of T if there is a non-zero function $f_\lambda \in L^\infty(X, \mathcal{B}, m)$ such that $f_\lambda(Tx) = \lambda f(x)$ a.e. m. We call any such f_λ an L^∞ eigenfunction of T corresponding to the eigenvalue λ. Since $\| f_\lambda \circ T \|_\infty = \| f_\lambda \|_\infty$ we have $| \lambda | = 1$. The collection $e(T)$ of L^∞ eigenvalues of T forms a subgroup of the circle group. Further

$$| f_\lambda(Tx) | = | \lambda | \, | f_\lambda(x) | = | f_\lambda(x) | \ a.e. \ m.$$

Since T is ergodic $| f_\lambda |$ is constant a.e. m. The function $\frac{f_\lambda}{|f|}$ is an eigenfunction of absolute value one, with eigenvalue λ.

11.2.   The collection $\mathcal{E}(T) = \mathcal{E}$ of all eigenfunctions of absolute value one is a group under pointwise multiplication. It is a closed subgroup of the group $\mathcal{U}$ of functions of absolute value one in $L^2(X, \mathcal{B}, m)$. The metric $d(f, g) = \| f - g \|_2$ is a complete metric on $\mathcal{E}$, invariant under group multiplication in $\mathcal{E}$. The collection of functions in $\mathcal{E}$ which are constant a.e. form a closed subgroup of $\mathcal{E}$ which is homeomorphically isomorphic to S^1.

11.3.   Consider the function $h : \mathcal{E} \to S^1$ defined by $h(f) = \overline{f}.foT$; $h(\mathcal{E}) = e(T)$, and h is continuous. Further, since T is ergodic, h is constant on cosets of S^1 in $\mathcal{E}$ and h assumes distinct values on distinct cosets. We may therefore view h as a one-one continuous function on the Polish group $\mathcal{E}/S^1$ onto $e(T)$. We denote this map from $\mathcal{E}/S^1$ onto $e(T)$ by $\tilde{h}$. Thus $e(T)$ is a Borel subset of S^1 since it is a one-one continuous image of the Polish space $\mathcal{E}/S^1$.

11.4.   We give $e(T)$ the topology under which the map $\tilde{h} : \mathcal{E}/S^1 \to e(T)$ becomes a homeomorphism. We denote this topology by $\mathcal{N}$. Note that $e(T)$ is

Polish in this topology and the injection map $i : i(x) = x$ from $e(T)$ into S^1 is continuous.

11.5. Suppose $\mathcal{N}_1$ is another topology on $e(T)$ under which $e(T)$ is a Polish group and the injection map i is continuous. Then $\mathcal{N}_1 = \mathcal{N}$. We see this as follows: Since the injection map is continuous under $\mathcal{N}_1$, the Borel structure generated by $\mathcal{N}_1$ agrees with the Borel structure inherited by $e(T)$ from S^1 which in turn is the Borel structure generated by $\mathcal{N}$. The injection $i : i(x) = x$ from $e(T)$ equipped with $\mathcal{N}$ onto $e(T)$ equipped with $\mathcal{N}_1$ is a one-one onto group isomorphism which is moreover a Borel map. Since a group homomorphism between Polish groups is continuous whenever it is Borel measurable [2], the map $i : e(T) \to e(T)$ is a homeomorphism between $e(T)$ equipped with $\mathcal{N}_1$ onto $e(T)$ equipped with $\mathcal{N}$.

11.6. Since S^1 is a closed subgroup of the Polish group $\mathcal{E}(T)$ we can choose a Borel set $B \subseteq \mathcal{E}(T)$ which intersects each S^1 coset in exactly one point. (This result due to J. Dixmier is now well known in descriptive set theory.) Thus, since $\mathcal{E}(T)/S^1$ is identified with $e(T)$, we see that there is a one-one Borel map $e(T) \to \mathcal{E}(T)$ with image B. An element of B corresponding to $t \in e(T)$ satisfies $\psi_t(Tx) = t\psi_t(x)$ a.e. m.

We may summarise the discussion so far as:

11.7 Theorem. *The group $e(T)$ is a Borel subset of S^1 and carries a unique Polish topology under which the injection map $i : x \to x$ from $e(T)$ into S^1 is continuous. (Usual topology on S^1). The Borel structure of $e(T)$ under this topology agrees with the Borel structure of $e(T)$ inherited from S^1. There is a Borel map $t \to \psi_t$ from $e(T)$ into $\mathcal{E}(T)$ such that for each t, ψ_t is an eigenfunction with eigenvalue t: $\psi_t(Tx) = t\psi_t(x)$ a.e. m.*

11.8. If m has a point mass at x (say), then m is supported on the orbit of x since T is ergodic. For any λ in S^1 the function f defined on the orbit of x by $f(T^n x) = \lambda^n$, $n \in \mathbb{Z}$, is an eigenfunction with eigenvalue λ (f may be defined equal to one outside the orbit of x). Thus if m is atomic then $e(T) = S^1$. We will see later that if $e(T) = S^1$ then m is atomic.

11.9 Definition. Let $Q \subseteq S^1$ be a countable infinite subgroup. Let $G = \hat{Q}_d = $ the compact dual of Q_d, where Q_d denotes the group Q with the discrete topology. Let $x_0 \in G$ be the element defined by $x_0(q) = q$ for all $q \in Q$. Let $\tau : G \to G$ be defined by $\tau x = x + x_0$. The system (G, τ) is called a compact group rotation.

11.10. It is to be noted that we do not have any fixed measure on G in mind (such as Haar measure) when we speak of compact group rotations, but we shall be concerned with measures on G which are quasi-invariant under τ. The next theorem is a non-singular version of the von Neumann-Halmos discrete spectrum theorem.

11.11 Theorem. *Assume that L^∞ eigenfunctions of T generate the σ-algebra $\mathcal{B}$ (modulo m-null sets). Then there is a compact group rotation (G, τ) and a finite measure ν on G, quasi-invariant and ergodic under τ, such that T and τ are isomorphic.*

Proof. Let $Q \subseteq e(T)$ be a countable subgroup dense in the topology of $e(T)$ described in **11.4**. Let $G = \hat{Q}_d$, the dual of Q with the discrete topology. Let τ be as in **11.9** for this G. For each $\lambda \in Q$ choose an eigenfunction f_λ and let $\mathcal{A}$ be the σ-algebra generated by all the f_λ's. Since any other eigenfunction is almost everywhere the limit of a sequence $c_n f_{\lambda_n}$, $n \in \mathbb{N}$, with $\lambda_n \in Q$, $c_n \in S^1$, we see that modulo m-null sets, $\mathcal{A}$ agrees with the σ-algebra generated by all eigenfunctions, which by assumption is $\mathcal{B}$ (mod m). We may therefore assume that the eigenfunctions f_λ, $\lambda \in Q$, generate $\mathcal{B}$ (mod m).

For $\lambda, \mu \in Q$, $f_{\lambda\mu}$ and $f_\lambda f_\mu$ are both L_∞ eigenfunctons with eigenvalue $\lambda\mu$. Therefore,

$$f_\lambda f_\mu = c(\lambda, \mu) f_{\lambda\mu} \ a.e.,$$

where $c(\lambda, \mu)$ is a complex number depending on the pair (λ, μ). Since Q is countable we can find a grand null set N such that for any $x \in X - N$, $f_\lambda(x) f_\mu(x) = c(\lambda, \mu) f_{\lambda\mu}(x)$, for all $\lambda, \mu \in Q$. Fix $x_0 \in X - N$ and set $\phi_\lambda = \frac{f_\lambda}{f_\lambda(x_0)}$, where $\lambda \in Q$. We have

$$c(\lambda, \mu) = \frac{f_\lambda(x_0) f_\mu(x_0)}{f_{\lambda\mu}(x_0)},$$

and for all $x \in X - N$, $\phi_\lambda(x)\phi_\mu(x) = \phi_{\lambda\mu}(x)$, for all $\lambda, \mu \in Q$.

Thus, since Q is countable, we may assume that the eigenfunctions f_λ, $\lambda \in Q$ are such that $f_{\lambda \cdot \mu} = f_\lambda \cdot f_\mu$. For each $x \in X$ the function $\lambda \to f_\lambda(x)$ is a character of Q. Further, since the f_λ's generate $\mathcal{B}$ (mod m) there is a T-invariant set $N \in \mathcal{B}$ of m measure zero such that for all distinct $x, y \in X - N$, $f_{(\cdot)}(x)$ and $f_{(\cdot)}(y)$ are distinct characters of Q. Therefore the map $S : X - N \to G = \hat{Q}_d$ given by $Sx = f_{(\cdot)}(x)$ is one-one and Borel measurable. (Borel measurable because for each λ in Q, $f_\lambda(x)$ is Borel measurable). Let us check that $S \circ T \circ S^{-1} = \tau$:

$$S \circ T \circ S^{-1} f_{(\cdot)}(x) = S \circ T(x) = f_{(\cdot)}(Tx) = (\cdot) f_{(\cdot)}(x) = \tau f_{(\cdot)}(x),$$

which verifies that $S \circ T \circ S^{-1} = \tau$. If we write $\nu = m \circ S^{-1}$, then ν is ergodic and quasi-invariant under τ. Clearly S preserves m since $\nu = m \circ S^{-1}$. The systems $(X, \mathcal{B}, m, T)$ and $(G, \mathcal{B}_G, \nu, \tau)$ are thus metrically isomorphic.

Quasi-Invariance of the Spectrum

11.12. We have seen in chapter 3 that the spectrum of V_ϕ can be different from that of U_T, both in spectral type and multiplicity. There is one characteristic, however, which remains the same for the spectrum of V_ϕ for all ϕ. It is the property that the spectral measure F of V_ϕ is quasi-invariant under $e(T)$ in the sense that F and F_q, $q \in e(T)$, are unitarily equivalent, where $F_q(A) = F(qA)$, $A \subseteq S^1$. To see this write for $q \in e(T)$

$$(W_q \xi)(x) = f_q(x)\xi(x), \ \xi \in L^2(X, \mathcal{B}, m),$$

where f_q is an eigenfunction of unit modulus with eigenvalue q. We note that with $V = V_\phi$,

$$V^n W_q = q^n W_q V^n, \ \text{or,} \ W_q^{-1} V^n W_q = q^n V^n, n \in \mathbb{Z}.$$

In terms of F this means that

$$W_q^{-1} F(A) W_q = F(qA), \ A \subseteq S^1, \tag{$*$}$$

so that F and F_q are unitarily equivalent.

Systems such as $(W_q, q \in Q, F)$ which satisfy $(*)$ are called systems of imprimitivity. They are mentioned in B. O. Koopman's original paper which connect spectral theory with dynamical systems. Their importance in more general settings was recognised by G. W. Mackey. We will discuss some generalities about such systems in the next chapter.

The Group $e(T)$ is σ-Compact

11.13 We will now show that the eigenvalue group $e(T)$ is a saturated subgroup of S^1. Moreover $e(T) = S^1$ if and only if m is discrete. For the definition of "saturated subgroup" see chapter **14**.

11.14 Proposition. *For any ergodic T the eigenvalue group $e(T)$ is a σ-compact subset in the usual topology of S^1.*

Proof. Let $E(T)$ denote the eigenfunctions of T of absolute value ≤ 1. Let $B(m)$ denote the unit ball in $L^2(X, \mathcal{B}, m)$ with weak the topology. It is compact and $B(m) - \{0\}$ is σ-compact in this topology. We note that $E(T)$ as a subset of $B(m) - \{0\}$ is closed in the weak topology, for suppose $\phi_n, n \in \mathbb{N}$, is a sequence of eigenfunctions in $E(T)$ with eigenvalues u_n, $n \in \mathbb{N}$, and suppose $\phi_n \to \phi$ *weakly*, $\phi \neq 0$. Then for all $f, g \in L^2(X, \mathcal{B}, m)$,

$$\int_X \phi_n f \cdot g dm \to \int_X \phi f \cdot g dm.$$

Now, if $U_T f = f \circ T^{-1} \sqrt{\frac{dm_{T^{-1}}}{dm}}$, then

$$
\begin{aligned}
\int_X (\phi_n \circ T) \cdot f \cdot g \, dm &= \int_X \phi_n \cdot (f \circ T^{-1}) \cdot g \circ T^{-1} dm \circ T^{-1} \\
&= \int_X \phi_n \cdot U_T f \cdot U_T g \, dm \\
&\to \int_X \phi \cdot U_T f \cdot U_T g \, dm \\
&= \int_X \phi \cdot f \circ T^{-1} \cdot g \circ T^{-1} dm \circ T^{-1} \\
&= \int_X \phi \circ T \cdot f \cdot g \, dm,
\end{aligned}
$$

This shows that $\phi_n \circ T \to \phi \circ T$ weakly. Now, over a subsequence the eigenvalues u_n converge to an element $u \in S^1$ which can be seen to be an eigenvalue with eigenfunction ϕ. (Indeed, therefore, u_n itself converges to u). Thus $E(T)$ is closed in $B(m) - \{0\}$. The map which assigns to any ϕ in $E(T)$ the eigenvalue of ϕ is a continuous function from $E(T)$ into S^1. Clearly the image of this map, which is $e(T)$, is σ-compact in the usual topology of S^1.

The Group $e(T)$ is Saturated

11.15. We know from **11.7** that there is a Borel map $t \to \psi_t$ from $e(T)$ into $\mathcal{E}(T)$ such that for all $t \in e(T)$, $\psi_t(Tx) = t\psi_t(x)$ a.e. m. However we do not know if the function $\psi(t, x)$ can be jointly measurable in t and x. We will show it can be chosen to be jointly measurable modulo the null sets of a measure. Let μ be a finite measure on $e(T)$. We will show that there is a jointly measurable function ϕ on $e(T) \times X$ such that for μ a.e. t, $\phi(t, Tx) = t\phi(t, x)$ a.e. m. To this end consider the complex measure on the Borel subsets of $e(T) \times X$ defined by

$$
\lambda(A \times B) = \int_A \left\{ \int_B \psi(t, x) dm \right\} d\mu.
$$

Let $\mathcal{D}_n = \{D_1^n, D_2^n, \ldots, D_{2^n}^n\}$ be a refining system of partitions in $e(T) \times X$ such that $\bigcup_{n=1}^{\infty} \mathcal{D}_n$ generates the σ-algebra of $e(T) \times X$. We define

$$
\phi_n(x, t) = \begin{cases} \frac{\lambda(D_k^n)}{\mu \times m(D_k^n)} & \text{if } (t, x) \in D_k^n, \mu \times m(D_k^n) \neq 0, \\ 1 & \text{otherwise.} \end{cases}
$$

Then each ϕ_n is jointly measurable and by the martingale convergence theorem $\phi_n \to \phi$ a.e. $\mu \times m$, where ϕ is jointly measurable and a version of $\frac{d\lambda}{d(\mu \times m)}$. For μ a.e t, $\phi(t, x) = \psi(t, x)$ m a.e. x. Since $\psi(t, .)$ is an eigenfunction with eigenvalue t, we see that for a.e. t (w.r.t. μ)

$$
\phi(t, Tx) = \psi(t, Tx) = t\psi(t, x) = t\phi(t, x) \text{ a.e. } m,
$$

which proves the claim.

11.16 Proposition. *A positive measure μ is concentrated on $e(T)$ if and only if there exists a Borel map $x \to \phi_x$ from X into $\mathcal{U}(\mu)$ (= functions in $L^2(S^1, \mu)$ of absolute value one) such that for a.e. x, $\phi_{Tx}(t) = t\phi_x(t)$ a.e. μ.*

Proof. If μ is concentrated on $e(T)$ then we choose a jointly measurable ϕ as above and set $\phi_x(t) = \phi(t, x)$. Conversely if such a Borel map $x \to \phi_x$ exists, then for a.e. t (w.r.t. μ), $\phi_{(\cdot)}(t)$ is an eigenfunction with eigenvalue t, whence μ is supported on $e(T)$.

11.17. We write $\overline{Z}_1(\mu)$ for the closure of the continuous characters in $L^2(S^1, \mu)$. Consider $S : \mathcal{U}(\mu) \to \mathcal{U}(\mu)$ defined by $(Sf)(t) = tf(t)$; S is continuous. Further for any character χ of S^1, $S\chi$ is also a character; $S\overline{Z}_1(\mu) = \overline{Z}_1(\mu)$. Since $\overline{Z}_1(\mu)$ is a closed subgroup of $\mathcal{U}(\mu)$, the coset space $\mathcal{U}(\mu)/\overline{Z}_1(\mu)$ admits a Borel cross-section C. Let q be the map on $\mathcal{U}(\mu)$ which sends $f \in \mathcal{U}(\mu)$ to the unique element $C \cap (f\overline{Z}_1(\mu))$. Then $p(f) = f \cdot (q(f))^{-1}$ belongs to $\overline{Z}_1(\mu)$ and

$$p(Sf) = (Sf)(q(Sf))^{-1} = Sf \cdot (q(f))^{-1} = Sp(f).$$

Thus we have proved:

11.18 Lemma. *For every positive measure μ on S^1 there is a Borel map $\mathcal{U}(\mu)$ onto $\overline{Z}_1(\mu)$ which commutes with S.*

Combining **11.16** and **11.18** we have:

11.19 Theorem. *A positive finite measure μ on S^1 is concentrated on $e(T)$ if and only if there exists a Borel map $x \to \xi_x$ from X to $\overline{Z}_1(\mu)$ such that for a.e. t (w.r.t. μ)*

$$\xi_{Tx}(t) = t\xi_x(t) \ a.e. \ m.$$

Proof. Suppose μ is concentrated on $e(T)$. Let $x \to \phi_x \in \mathcal{U}(\mu)$ be a function as in **11.16** and let $\xi_x = p(\phi_x)$. Now $\phi_{Tx} = S(\phi_x)$ (from the definition of S). Hence

$$\xi_{Tx} = p(\phi_{Tx}) = p(S\phi_x) = Sp(\phi_x) = S\xi_x \ a.e. \ m.$$

Thus for a.e. t (w.r.t. μ), $\xi_{Tx}(t) = t\xi_x(t)$ a.e.m. Conversely if the function ξ as postulated exists, then for a.e. t (w.r.t. μ), $\xi_{(\cdot)}(t)$ is an eigenfunction of T with eigenvalue t so that μ a.e. t is in $e(T)$, i.e., μ is supported on $e(T)$.

We now prove the claim made in **11.13**.

11.20 Theorem. *If T is ergodic then the group $e(T)$ of L^∞ eigenvalues of T is a saturated subgroup of S^1. Further $e(T) = S^1$ if and only if m is discrete.*

Proof. For the definition of saturated subgroup see chapter **14**. We will show that every measure which sticks to a measure concentrated on $e(T)$ is itself concentrated on $e(T)$. To this end let μ be a measure on $e(T)$ and let ν be a measure on S^1 which sticks to μ. This means that whenever $\hat{\mu}(n_k) \to \| \mu \|$, the Fourier coefficients $\hat{\nu}(n_k) \to \| \nu \|$. Equivalently there exists a continuous group homomorphism $h : \overline{Z}_1(\mu) \to \overline{Z}_1(\nu)$ which maps the characters χ_n in $\overline{Z}_1(\mu)$ to χ_n in $\overline{Z}_1(\nu)$. We also have $hS = Sh$. Let $x \to \xi_x$ be the map provided by theorem **11.19**. Then

$$h(\xi_{Tx}) = h(S\xi_x) = Sh(\xi_x), a.e.\ m.$$

Write $\xi'_x = h(\xi_x)$, then the function $x \to \xi'_x$ from X to $\overline{Z}_1(\nu)$ has the property that for ν a.e. t, $\xi'_{(.)}(t)$ is an eigenfunction of T with eigenvalue t, which means that ν is concentrated on $e(T)$. Thus $e(T)$ is a saturated subgroup. Suppose now that $e(T) = S^1$. We can apply theorem **11.19** with μ equal to the Lebesgue measure. Since $\overline{Z}_1(\text{Lebesgue}) = $ all continuous characters on S^1, we may identify $\overline{Z}_1(\mu)$ with the group of integers $\mathbb{Z}$. The map $x \to \xi_x \in \mathbb{Z}$ is such that T on X and shift $n \to n+1$ on $\mathbb{Z}$ commute; whence set $W = \{x : \xi_x = 1\}$ is a wandering set with $\cup_{n=-\infty}^{\infty} T^n W = X$ (mod m). Thus T is dissipative, and since it is ergodic, m is discrete. Conversely, if m is discrete we already know that $e(T) = S^1$.

11.21 Corollary. *If $e(T) \neq S^1$ then $e(T)$ is a weak Dirichlet set.*

Proof. Since $e(T)$ is σ-compact and saturated, it is a weak Dirichlet set whenever it is not all of S^1.

11.22. The discussion of **11.1–11.7** is an elaboration of facts about $e(T)$ mentioned in C. C. Moore and K. Schmidt [4], K. Schmidt [5] who also shows that $e(T)$ is a weak Dirichlet set whenever $e(T) \neq S^1$. For the discussion **11.1–11.12**, see J. Aaronson and M. Nadkarni [1]). For the discussion **11.13–11.21**, we have drawn on B. Host, J.-F. Méla and F. Parreau [3]).

Chapter 12

Generalities on Systems of Imprimitivity

12.1. Consider a measure preserving system $(X, \mathcal{B}, m, T)$ and the spectral E measure defined on $(X, \mathcal{B})$ by:

$$E(A)f = 1_A f, \quad f \in L^2(X, \mathcal{B}, m).$$

We note that U_T, the unitary operator associated with T (defined by $U_T f = f \circ T^{-1}$), satisfies with E the relation

$$U_T^n E(A) U_T^{-n} = E(T^n A), A \in \mathcal{B}.$$

In this chapter we consider such a relation in a general setting. We will discuss spectral measures on a Borel space $(X, \mathcal{B})$ and groups of unitary operators which together satisfy a relation similar to above. Such systems are called systems of imprimitivity, (see V. S. Varadarajan [5]).

Spectral Measures and Group Actions

12.2. Let E be a spectral measure on $(X, \mathcal{B})$ acting in a complex separable Hilbert space $\mathcal{H}$. Let G be a family of one-one maps of X onto X such that for all $g \in G$, $g\mathcal{B} = g^{-1}\mathcal{B} = \mathcal{B}$. For the time being we need not assume that G is a group. Let E_g be defined by $E_g(B) = E(gB)$, $g \in G$. Let $[\nu_\infty], [\nu_1], [\nu_2], \ldots$ be the mutually singular measure classes associated with E as per the second form of the Hahn-Hellinger theorem. Then $[\nu_{\infty g}], [\nu_{1g}], [\nu_{2g}], \ldots$ are the mutually singular measure classes similarly associated with E_g. (Here ν_g denotes the measure $\nu_g(B) = \nu(gB), B \in \mathcal{B}$). Now suppose that for each $g \in G$, a unitary operator V_g is defined on $\mathcal{H}$ such that

$$V_g E(B) V_g^{-1} = E(gB), \quad \text{for all } B \in \mathcal{B}. \tag{1}$$

This means that E and E_g are unitarily equivalent for all $g \in G$. By the Hahn-Hellinger theorem (second form) we conclude that for each $g \in G$, and each i,

$[\nu_i] = [\nu_{ig}]$. If we look at the subspace

$$\mathcal{H}_i = \{x \in \mathcal{H} : \mu_x \ll \nu_i\}.$$

where μ_x is the measure defined by $\mu_x(A) = (E(A)x, x)$, then $\mathcal{H}_i$ is invariant under all V_g and all $E(A), A \in \mathcal{B}$. If E_i denotes the restriction of E to $\mathcal{H}_i$ and $V_{g,i}$ that of V_g, then we have the relation

$$V_{g,i}E_i(A)V_{g,i}^{-1} = E_i(gA)$$

satisfied for all $A \in \mathcal{B}$. The system $(V_g, g \in G, E)$ satisfying (1) may therefore be studied by first looking at the case when E has uniform multiplicity.

12.3. Assume, then, that E has uniform multiplicity $k \le \aleph_0$ and that (1) holds. Then only ν_k is non-zero. We denote this measure by ν. Let $K = \mathbb{C}^k$ if $k < \aleph_0$ and $K = l^2$, if $k = \aleph_0$. Now by the Hahn-Hellinger theorem there is an isometry S from $\mathcal{H}$ onto $L^2(X, \nu, K)$ such that for all f in the latter space:

$$(SE(B)S^{-1}f)(x) = 1_B(x)f(x), B \in \mathcal{B}. \tag{2}$$

Let $\tilde{V}_g = SV_gS^{-1}$. We will now compute $\tilde{V}_g$ in more detail in order to arrive at (6) below. Now (1) holds, therefore

$$\tilde{V}_g^{-1}\tilde{E}(B)\tilde{V}_g = \tilde{E}(gB), \tag{3}$$

where $\tilde{E}(B)f = 1_Bf, B \in \mathcal{B}$.

Let U_g be defined by

$$(U_gf)(x) = \left(\frac{d\nu_g}{d\nu}(x)\right)^{1/2} f(gx), \ f \in L^2(X, \nu, K).$$

Then

$$U_g^{-1}f = \left(\frac{d\nu_{g^{-1}}}{d\nu}(x)\right)^{1/2} f(g^{-1}x),$$

$$
\begin{aligned}
(U_g^{-1}\tilde{E}(B)U_gf)(x) &= U_g^{-1}(1_B(x))\left(\frac{d\nu_g}{d\nu}(x)\right)^{1/2} f(gx) \\
&= 1_B(g^{-1}x)\left(\frac{d\nu_{g^{-1}}}{d\nu}(x)\right)^{1/2}\left(\frac{d\nu_g}{d\nu}(g^{-1}x)\right)^{1/2} f(x) \\
&= (\tilde{E}_g(B)f)(x),
\end{aligned}
$$

(we have used here the fact that

$$\frac{d\nu_{g^{-1}}}{d\nu}(x)\left(\frac{d\nu_g}{d\nu}(g^{-1}x)\right) = 1 \ \text{a.e.).}$$

Thus we have:
$$U_g^{-1}\tilde{E}(B)U_g = \tilde{E}(gB), \quad \text{for all } B \in \mathcal{B}. \tag{4}$$

Now consider $W_g = \tilde{V}_g U_g^{-1}$. For $f \in L^2(X, \nu, K)$,

$$
\begin{aligned}
(W_g^{-1}1_B W_g f)(x) &= (U_g \tilde{V}_g^{-1} 1_B \tilde{V}_g U_g^{-1} f)(x) \\
&= (U_g 1_{gB} U_g^{-1} f)(x) \quad \text{by (3)} \\
&= (U_g U_g^{-1} 1_B U_g U_g^{-1} f)(x) \quad \text{by (4)} \\
&= 1_B(x) f(x) \text{ a.e.}
\end{aligned}
$$

Thus
$$1_B W_g = W_g 1_B,$$

for all $B \in \mathcal{B}$. Since the unitary operator W_g commutes with all multiplications, there is a measurable function $\tilde{W}_g$ on $(X, \mathcal{B})$ whose values are unitary operators on K such that
$$(W_g f)(x) = \tilde{W}_g(x) f(x), \quad \text{a.e. } \nu.$$

Since $K = \mathbb{C}^k$ or l^2 depending on whether $k < \aleph_0$ or $k = \aleph_0$, we may assume that W is indeed a $k \times k$ unitary matrix valued function with measurable entries. Now $W_g = \tilde{V}_g U_g^{-1}$ hence $\tilde{V}_g = W_g U_g$, or

$$(\tilde{V}_g f)(x) = \tilde{W}_g(x) \left(\frac{d\nu_g}{d\nu}(x) \right)^{1/2} f(gx) \text{ a.e.} \tag{6}$$

for all $f \in L^2(X, \nu, K)$. This yields the desired form of $\tilde{V}_g$.

12.4. Let S' be another isometry which accomplishes (2), then there will be corresponding $\tilde{V}_g$, $\tilde{W}_g$, denoted by $\tilde{V}'_g$, $\tilde{W}'_g$ respectively which will be related as in (6). We now compute the relation between $\tilde{W}_g$ and $\tilde{W}'_g$. Let $\tilde{V}'_g = S' V_g S'^{-1}$. Now $S'S^{-1}$ commutes with multiplication by 1_B for all $B \in \mathcal{B}$, hence it is of the form

$$(S'S^{-1}f)(x) = A(x)f(x),$$

where A is a $k \times k$ matrix with measurable entries. Further,

$$
\begin{aligned}
(\tilde{V}'_g f)(x) &= (S' V_g S'^{-1} f)(x) \\
&= (S'S^{-1} \tilde{V}_g SS'^{-1} f)(x) \\
&= A(x)\tilde{W}_g(x)A^{-1}(gx) \left(\frac{d\nu_g}{d\nu}(x) \right)^{1/2} f(gx). \tag{7}
\end{aligned}
$$

Now

$$(V'_g f)(x) = \tilde{W}'_g(x) \left(\frac{d\nu_g}{d\nu}(x) \right)^{1/2} f(gx). \tag{8}$$

Comparing (7) and (8), since these hold for all f, we get

$$\tilde{W}'_g(x) = A(x)\tilde{W}_g(x)A^{-1}(gx)$$

which is the desired relation between $\tilde{W}'_g$ and $\tilde{W}_g$.

12.5. Assume in the rest of this chapter that G is a group of Borel automorphisms on $(X, \mathcal{B})$ and that $V_g, g \in G$, is a group of unitary operators on $\mathcal{H}$ satisfying with E the relation (1). Then, with $\tilde{V}_g = SV_gS^{-1}$, we have

$$(\tilde{V}_{gh}f)(x) = \tilde{W}_{gh}(x)\left(\frac{d\nu_{gh}}{d\nu}(x)\right)^{1/2} f(ghx) \text{ a.e.} \tag{9}$$

And also

$$(\tilde{V}_{gh}f)(x) = (\tilde{V}_g(\tilde{V}_hf))(x) \text{ a.e.}$$

$$= \tilde{W}_g(x)\tilde{W}_h(gx)\left(\frac{d\nu_g}{d\nu}(x)\right)^{1/2}\left(\frac{d\nu_h}{d\nu}(gx)\right)^{1/2} f(ghx). \tag{10}$$

Now

$$\frac{d\nu_{gh}}{d\nu}(x) = \frac{d\nu_g}{d\nu}\frac{d\nu_h}{d\nu}(gx) \text{ a.e.} \tag{11}$$

From this and the fact that right hand sides of (9) and (10) agree for all f we get

$$\tilde{W}_{gh}(x) = \tilde{W}_g(x)\tilde{W}_h(gx) \text{ a.e.}$$

Considerations of **12.4** and **12.5** allow us to formulate the definitions **12.6**, **12.7** and theorem **12.8**.

Cocycles; Systems of Imprimitivity

12.6 Definition. Let ν be a finite measure on $(X, \mathcal{B})$, quasi-invariant under a group G of Borel automorphisms on $(X, \mathcal{B})$. A function C on $G \times X$ (whose values are unitary operators on a complex separable Hilbert space K) is called cocycle if

  (i) for each g, $C(g, .)$ is weakly measurable,
  (ii) $C(gh, x) = C(g, x) \cdot C(h, gx)$ a.e. ν.

Two cocycles C and D on $G \times X$ (with respect to the same measure class ν) are said to be cohomologous if there exists a weakly measurable unitary operator valued function A such that for all $g \in G$,

$$D(g, x) = A(x) \cdot C(g, x) \cdot A^{-1}(gx) \text{ a.e. } \nu. \tag{12}$$

The ν null set where (12) fails to hold may depends on g. A cocycle C is said to a coboundary if it is of the form $C(g, x) = A(x) \cdot A^{-1}(gx)$ where A is a weakly measurable unitary operator valued function.

Remark. The identity **(ii)** above is referred to as the cocycle identity and any function on $G \times X$ which satisfies such a relation is said to satisfy the cocycle identity. Thus, in view of (11), $\frac{d\nu_g}{d\nu}(x)$ as a function on $G \times X$ satisfies the cocycle identity.

12.7 Definition. Let $(X, \mathcal{B})$, G, be as in **12.6**. Let E be a spectral measure on $\mathcal{B}$ acting in a complex separable Hilbert space $\mathcal{H}$, and, let $V_g, g \in G$, be a group of unitary operators on $\mathcal{H}$ which satisfies with E the relation

$$V^{-1}EV_g = E_g \quad \text{for all } g \in G.$$

We then say that $(V_g, g \in G, E)$ is a system of imprimitivity based on (G, X) acting in $\mathcal{H}$. Taking into account the considerations of sections **12.2-12.6** we have:

12.8 Theorem. *Let $(V_g, g \in G, E)$ be a system of imprimitivity based on (G, X) acting in $\mathcal{H}$. Let E have uniform multiplicity k and let K be a k dimensional Hilbert space. Then there exists a finite measure ν quasi-invariant under G and an invertible isometry*

$$S : \mathcal{H} \leftrightarrow L^2(X, \nu, K)$$

such that for all $f \in L^2(X, \nu, K)$ and $B \in \mathcal{B}$,

$$SE(B)S^{-1}f = 1_B f,$$

$$(SV_g S^{-1}f)(x) = \tilde{W}(g, x) \left(\frac{d\nu_g}{d\nu}(x)\right)^{1/2} f(gx), \quad g \in G,$$

where $\tilde{W}(g, x)$ is a $G \times X$ cocycle relative to ν taking values in $\mathcal{U}(K)$. If ν' is another such measure, S' another such isometry, and $\tilde{W}'$ the corresponding cocycle, then ν, ν' have the same measure class and the cocycles $\tilde{W}$, $\tilde{W}'$ are cohomologous.

Here $\mathcal{U}(K)$ is the class of unitary operators in K.

12.9. A system of imprimitivity in which E is not of uniform multiplicity is, in an obvious sense, a direct sum of systems of imprimitivity of uniform multiplicity.

12.10 Definition. Two systems of imprimitivity $(V_g, g \in G, E)$, $(V_g', g \in G, E')$ based on (X, G) and acting in separable Hilbert spaces $\mathcal{H}$ and $\mathcal{H}'$ respectively are said to be equivalent if there is an isometry S from $\mathcal{H}$ onto $\mathcal{H}'$ such that

(i) $SE(B)S^{-1} = E'$ for all $B \in \mathcal{B}$,
(ii) $SV_g S^{-1} = V_g'$ for all $g \in G$.

12.11. Let λ be a σ-finite measure on $\mathcal{B}$, quasi-invariant under G. Let K be a complex separable Hilbert space and assume that $L^2(X, \lambda, K)$ is separable.

Let $\mathcal{U}(K)$ be equipped with the Borel structure stemming from the topology of weak convergence. Let A be a $G \times X$ cocycle taking values in $\mathcal{U}(K)$. With these objects we can define a system of imprimitivity acting in $L^2(X, \lambda, K)$ as follows: for $f \in L^2(X, \lambda, K)$, define

(i) $E(B)f = 1_B f, B \in \mathcal{B}$,

(ii) $(V_g f)(x) = A(g, x)\left(\frac{d\lambda_g}{d\lambda}(x)\right)^{1/2} f(gx), \quad g \in G$.

Such a system is called a concrete system of imprimitivity of dimension k, where k is the dimension of K. If $(V_g', g \in G, F)$ is another concrete system of imprimitivity based on (X, G) and acting in $L^2(X, \nu, K)$ with associated cocycle A', then the two systems are unitarily equivalent if and only if (i) λ and ν have the same class of null sets, and (ii) the cocycles A and A' are cohomologous modulo the measure class. We may restate theorem **12.8** briefly as:

12.12 Theorem. *Every system of imprimitivity for which the associated spectral measure has uniform multiplicity k, is unitarily equivalent to a concrete system of imprimitivity of dimension k.*

Irreducible Systems of Imprimitivity

12.13 Definition. A system of imprimitivity $(U_g, g \in G, E)$ acting in $\mathcal{H}$ is said to be irreducible if the only subspaces of $\mathcal{H}$ invariant under all $U_g, g \in G$, and all $E(A), A \in \mathcal{B}$, are the trivial ones.

12.14. If $(U_g, g \in G, E)$ is an irreducible system of imprimitivity then

(i) E has uniform multiplicity,

(ii) Any quasi-invariant measure λ which has the same null sets as E is ergodic under the G action in the sense that for any G-invariant Borel set A, either $\lambda(A) = 0$ or $\lambda(X - A) = 0$.

Now (i) is obvious. We see (ii) as follows: If λ is not ergodic under G, then there is a Borel set A invariant under G such that $\lambda(A) > 0$ and $\lambda(X - A) > 0$. Then the subspaces $\mathcal{H}_1 = \{x \in \mathcal{H} : \mu_x \ll \lambda \mid_A\}$ and $\mathcal{H}_2 = \{x \in \mathcal{H} : \mu_x \ll \lambda \mid_{X-A}\}$ are non-trivial and invariant under all $U_g, g \in G$, and all $E(B), B \in \mathcal{B}$. This contradicts the irreducibility of the system.

(iii) If the maximal spectral type λ of E is ergodic with respect to G and if E has multiplicity one then the system $(U_g, g \in G, E)$ is irreducible. An example of an irreducible system of imprimitivity with the multiplicity of the associated E greater than one will be given in the next chapter. Note that uniform multiplicity and ergodicity of E are necessary but not sufficient conditions for irreducibility of the system. Any non-trivial concrete system of imprimitivity in which the cocycle is the identity matrix of dimension greater than one is reducible irrespective of the measure.

Transitive Systems

12.15 Remark. Assume that $(X, \mathcal{B})$ is a standard Borel space and G is a locally compact second countable group which acts on X so that the map $(g, x) \to gx$ from $G \times X$ to X is jointly measurable. Let $(U_g, g \in G, E)$ be an irreducible system of imprimitivity, where $U_g, g \in G$, is a continuous representation of G. If we represent this system as a concrete system of imprimitivity and if A and μ are the cocycle and the measure associated with the concrete representation of $(U_g, g \in G, E)$, then A can be chosen to be $G \times X$ measurable; also a version of $\frac{d\mu_g}{d\mu}$ can be chosen to be $G \times X$ measurable. The action is said to be transitive if there is an $x_0 \in X$ such that μ is carried on Gx_0; the system of imprimitivity is then called a transitive system of imprimitivity. Such systems are important in the representation theory of non-commutative locally compact groups (see G. W. Mackey [2], V. S. Varadarajan [5]).

Transitive Systems on $\mathbb{R}$

12.16. We discuss irreducible transitive systems on $\mathbb{R}$. Let $G = \mathbb{R}$ equipped with the usual topology and let $\mathbb{R}$ act on $\mathbb{R}$ by $(t, x) \to x + t$. On $\mathcal{H} = L^2(\mathbb{R}, l)$, where l denotes the Lebesgue measure, define

$$U_t f = f_t = \text{ the translate of } f \text{ by } t,$$

$$E(A)f = 1_A f, \quad A \in \mathcal{B}_{\mathbb{R}}.$$

Then $(U_t, t \in \mathbb{R}, E)$ is an irreducible system of imprimitivity. To see this note that a subspace invariant under all $E(A), A \in \mathcal{B}_{\mathbb{R}}$, consists of functions in $L^2(\mathbb{R}, l)$ which vanish outside a fixed set $\Lambda \in \mathcal{B}_{\mathbb{R}}$. If such a subspace is to be invariant under all $U_t, t \in \mathbb{R}$, then it must be either $L^2(\mathbb{R}, l)$ or $\{0\}$. Further, if for this action of $\mathbb{R}$, $(U'_t, t \in \mathbb{R}, E')$ is another irreducible system of imprimitivity (where $U'_t, t \in \mathbb{R}$, is a continuous group of unitary operators) then $(U'_t, t \in \mathbb{R}, E')$ is unitarily equivalent to $(U_t, t \in \mathbb{R}, E)$.

Let

$$V_t = \int_{-\infty}^{\infty} e^{itx} dE,$$

then $(V_t f)(x) = e^{itx} f(x)$ a.e. Further

$$U_s V_t = e^{its} V_t U_s, \quad V_{-t} U_s V_t = e^{its} U_s.$$

If $U_s = \int_{-\infty}^{\infty} e^{-isx} dF(x)$ is the spectral resolution of $U_s, s \in \mathbb{R}$, then

$$V_{-t} U_s V_t = \int_{-\infty}^{\infty} e^{-isx} V_{-t} dF V_t.$$

Also

$$V_{-t}U_sV_t = e^{its} \int_{-\infty}^{\infty} e^{-isx}dF = \int_{-\infty}^{\infty} e^{-is(x-t)}dF = \int_{-\infty}^{\infty} e^{-isu}dF_t,$$

where $F_t(A) = F(A+t)$. We thus have

$$V_{-t}F(A)V_t = F(A+t), \text{ for all } A \in \mathcal{B}_{\mathbb{R}} \text{ and } t \in \mathbb{R},$$

so that $(V_t, t \in \mathbb{R}, F)$ is a new system of imprimitivity arising out of $(U_s, s \in \mathbb{R}, E)$.

Dual systems of imprimitivity such as $(U_s, s \in \mathbb{R}, E), (V_t, t \in \mathbb{R}, F)$, but which are non-transitive, arise naturally in the work of Helson and Lowdenslager [2] in their generalisation of the H^2 theory and prediction theory to more general groups. If the transitive action of $\mathbb{R}$ on $\mathbb{R}$ is replaced by that of a countable dense subgroup of $\mathbb{R}$, a theory of compact groups with ordered duals emerges which connects the H^2 theory with the spectral theory of strictly ergodic actions (see Helson [1]). The consideration of objects such as eigenvalues of non-singular automorphisms, the group $H(\mu)$ of chapter 11 and the study of cocycles for strictly ergodic actions was stimulated by problems arising in this theory. In the next chapter we will give a discussion of dual systems of imprimitivity.

Chapter 13

Dual Systems of Imprimitivity

Compact Group Rotations; Dual Systems of Imprimitivity

13.1. We now discuss systems of imprimitivity based on compact group rotations. In this case there are naturally arising dual systems of imprimitivity and the two together yield considerable spectral information. First we recall the definition of a compact group rotation: Let $Q \subseteq S^1$ be a countable infinite group. Let $G = \hat{Q}_d$ be the compact dual of Q_d, where Q_d is the group Q with the discrete topology. Let $x_0 \in G$ be the element defined by $x_0(q) = q$ for all $q \in Q$. Let $\tau : G \to G$ be defined by $\tau x = x + x_0, x \in G$. Then the system (G, τ) is called a compact group rotation.

13.2 Let (G, τ) be a compact group rotation. We view (G, τ) alternatively as the $\mathbb{Z}$ space $(\mathbb{Z}, G)$ where the $\mathbb{Z}$ action on G is given by $n(x) = \tau^n x, x \in G, n \in \mathbb{Z}$. We speak of the system of imprimitivity being based on (G, τ) rather than on $(\mathbb{Z}, G)$. Let $(V^n, n \in \mathbb{Z}, E)$ be a system of imprimitivity based on (G, τ) acting in a complex separable Hilbert space $\mathcal{H}$. We then have

$$V^{-n} E(B) V^n = E(\tau^n B), B \subseteq G, n \in \mathbb{Z}. \tag{1}$$

Let

$$U_q = \int_G \chi_q^{-1} dE, \quad q \in Q,$$

where χ_q denotes the character on G corresponding to $q \in Q$. Write $E_{\tau^n}(A) = E(\tau^n A)$. Now,

$$
\begin{aligned}
V^{-n} U_q V^n &= \int_G \chi_q^{-1} V^{-n} dE V^n = \int_G \chi_q^{-1} dE_{\tau^n} \quad \text{by (1)}, \\
&= \int_G \chi_q^{-1}(y - nx_0) dE = q^n U_q.
\end{aligned}
$$

We thus have

$$V^{-n} U_q V^n = q^n U_q, \qquad U_q V^n U_q^{-1} = q^n V^n. \tag{2}$$

If

$$V^n = \int_{S^1} z^n dF$$

is the spectral resolution of $V^n, n \in \mathbb{Z}$, then we have for all $q \in Q$,

$$U_q V^n U_q^{-1} = \int_{S^1} z^n U_q dF U_q^{-1}, \tag{3}$$

$$U_q V^n U_q^{-1} = q^n V^n = \int_{S^1} (qz)^n dF = \int_{S^1} z^n dF_{q^{-1}} \tag{4}$$

where $F_{q^{-1}}(A) = F(q^{-1}A), A \subseteq S^1$. We now compare the right hand sides of (3) and (4) and obtain

$$U_q F(A) U_q^{-1} = F(q^{-1}A), \quad A \subseteq S, \quad q \in Q.$$

Thus $(U_q, q \in Q, F)$ is a system of imprimitivity naturally arising from $(V^n, n \in \mathbb{Z}, E)$. We call $(U_q, q \in Q, F)$ the dual of the system $(V^n, n \in \mathbb{Z}, E)$. Thus we have a pair of systems of imprimitivity each dual of the other. We may write $(V^n, n \in \mathbb{Z}, E)^\wedge = (U_q, q \in Q, F)$.

Irreducible Dual Systems; Examples

13.3. It is clear that a closed subspace of the Hilbert space $\mathcal{H}$ is invariant under all $V^n, n \in \mathbb{Z}$, and all $E(B), B \subseteq G$, if and only if it is invariant under all $F(A), A \subseteq S^1$, and all $U_q, q \in Q$. The system $(V^n, n \in \mathbb{Z}, E)$ is irreducible if and only if the system $(U_q, q \in Q, F)$ is irreducible.

13.4. We now discuss such irreducible dual pairs, which moreover we consider in their concrete form. Let m be a probability measure on $\mathcal{B}$ ($=$ Borel subsets of G) having the same null sets as E. Since $(V^n, n \in \mathbb{Z}, E)$ is irreducible, m is quasi-invariant and ergodic under τ. Let E have multiplicity k, and let $\mathcal{H}_k$ denote a Hilbert space of dimension k. We may assume that the irreducible system $(V^n, n \in \mathbb{Z}, E)$ acts on $L^2(G, m, \mathcal{H}_k)$ in the following manner:

$$E(B)f = 1_B f, B \subseteq G, \ f \in L^2(G, m, \mathcal{H}_k),$$

$$(V^n f)(x) = A(n, x) \left(\frac{dm_{\tau^n}}{dm}\right)^{1/2} (x) f(\tau^n x), \ n \in \mathbb{Z}.$$

where A is a $\mathbb{Z} \times G$ measurable cocycle (relative to m) whose values are unitary operators on $\mathcal{H}_k$.

Let $(U_q, q \in Q, F)$ be the dual of $(V^n, n \in \mathbb{Z}, E)$. Let μ be a probability measure on S^1 having the same null sets as F. Irreducibility of $(U_q, q \in Q, F)$ again implies that μ is quasi-invariant and ergodic under the Q action ($x \to qx, \ q \in Q$), on S^1. Let F have multiplicity l. Let $\mathcal{H}_l$ be a Hilbert space of dimension l. The system $(U_q, q \in Q, F)$ is defined on $L^2(G, m, \mathcal{H}_k)$. However by the Hahn-Hellinger theorem there is an isometry S from $L^2(G, m, \mathcal{H}_k)$ onto $L^2(S^1, \mu, \mathcal{H}_l)$ such that

$$SF(B)S^{-1}g = 1_B g, \ B \subseteq S^1, \ g \in L^2(S^1, \mu, \mathcal{H}_l),$$

$$(SU_q S^{-1}g)(x) = C(q, x) \left(\frac{d\mu_q}{d\mu}\right)^{1/2}(x) g(qx), \ q \in Q,$$

where $C(q, x)$ is a $Q \times S^1$ measurable cocycle whose values are unitary operators on $\mathcal{H}_l$. The measure class of m and the cohomology class of A together determine the measure class of μ and the cohomology class of C and vice versa. We may thus write

$$(m, A)^\wedge = (\mu, C), (\mu, C)^\wedge = (m, A).$$

We discuss some of special cases.

13.5. Consider the case when $k = 1$ and $A = 1$. We then get an irreducible system since E has multiplicity one and m is ergodic. The dual system $(U_q, q \in Q, F)$ is then irreducible. The maximal spectral type of F is quasi-invariant and ergodic under Q and F has uniform multiplicity. It is not known whether the multiplicity of F is always one or can be higher in this case. We write ν for the maximal spectral type of F.

13.6. We consider example **3.11** in the light of the above discussion. Let $Q = \{\alpha^n, n \in \mathbb{Z}\}$, where $\alpha \in S^1$ is such that Q is dense in S^1. Since Q_d is isomorphic to $\mathbb{Z}$, the dual of Q_d is S^1. If $x \in S^1$ is regarded as a character on Q, then $x(\alpha^n) = x^n$. Now $x \in S^1$ which acts on Q_d as the identity map, i.e., for which $x(\alpha^n) = \alpha^n$ for all $n \in \mathbb{Z}$, is indeed $x = \alpha$. Thus $G = \hat{Q}_d = S^1$, $\tau x = \alpha x$, and (G, τ) is the system dual to (S^1, Q); it is (S^1, Q) itself.

13.6. (continued). Let m be the normalised Haar measure on S^1 and $A(n, x) = \beta^n$, where $\beta \in S^1$. $A(\cdot, \cdot)$ is a $Q \times S^1$ coboundary if and only if β is of the form α^n for some n. In this case E and V^n, $n \in \mathbb{Z}$, act on $L^2(S^1, m)$ as follows:

$$E(A)f = 1_A f, \ A \subseteq S^1, \ f \in L^2(S^1, m),$$

$$(V^n f)(x) = \beta^n f(\alpha^n x), n \in \mathbb{Z}.$$

The multiplicity of E is one and so is that of F, since V admits a complete orthonormal set of eigenvectors, each of multiplicity one. The eigenvalues of V are $\beta\alpha^k$ with characters $\chi_k, k \in \mathbb{Z}$, as the corresponding eigenvectors. The maximal spectral type μ of F is discrete and supported on $\{\beta\alpha^k\}, k \in \mathbb{Z}$. The cocycle C is a coboundary because the Q action on S^1 is transitive and free with respect to μ.

13.7. We return to the case of the general compact group rotation (G, τ) of **13.2.** Let m be quasi-invariant and ergodic under τ. Assume that $A(n, x) = \beta^n$

for some $\beta \in S^1$. Then μ is in the measure class of γ_β, $(\gamma_\beta(B) = \gamma(\beta B))$ where γ is the maximal spectral type of F when $A(n,x) = 1$ for all n, x. In addition, if m is equivalent to Haar measure, then γ is a discrete measure on Q_d. The cocycle C, for the case when $m = $ Haar measure on G, is therefore a coboundary.

13.8. Call a cocycle A trivial if it is constant in x for each fixed n. It is of the form $A(n,x) = \beta^n$ for some $\beta \in S^1$. We have seen above that if A is trivial and $m = $ Haar measure on G, then $(A, m)^\wedge = (C, \mu)$, where μ is a discrete measure on βQ and C is a coboundary.

13.9. From **13.7** it may be natural to surmise that whenever μ is quasi-invariant and ergodic and C is a scalar coboundary, the cocycle A is trivial. This however is false. We see this as follows: Suppose C is a coboundary and $A(n,x) = \beta^n$ for some $\beta \in S^1$. Since C is a scalar coboundary, we may assume that $C = 1$. Next let $\gamma = \mu_{\beta^{-1}}$. Then the group action $x \to qx$, $q \in Q$, $x \in S^1$, when considered as two separate actions, one with respect to μ and the other with respect γ, are metrically isomorphic since γ is a translate of μ. Hence the group of unitary operators

$$(U_q f)(x) = \left(\frac{d\mu_q}{d\mu}\right)^{1/2}(x)f(qx), \quad f \in L^2(S^1, \mu), \quad q \in Q,$$

$$(W_q f)(x) = \left(\frac{d\gamma_q}{d\gamma}\right)^{1/2}(x)f(qx), f \in L^2(S^1, \gamma), q \in Q,$$

are unitarily equivalent. The system of imprimitivity dual to $(W_q, q \in Q, F_1)$, $(F_1(A) = 1_A f,\ A \subseteq S^1,\ f \in L^2(S^1, \gamma))$, is described by the measure class of m and the cocycle $A = 1$. The spectral measure F_1 is therefore unitarily equivalent to the spectral measure of

$$(V_1^n f)(x) = \left(\frac{dm_{\tau^n}}{dm}\right)^{1/2}(x)f(\tau^n x), \quad f \in L^2(G, m), \quad n \in \mathbb{Z},$$

which is symmetric, (Chapter 3). Thus F_1 is symmetric, so that $\gamma(A) > 0$ if and only if $\gamma(A^{-1}) > 0$. Dually we can surmise that if m is ergodic and quasi-invariant under τ then some translate of m is symmetric in the sense that $m(A) > 0$ if and only if $m(-A) > 0$. This, however, is false. For example if one takes m on $\{0,1\}^\mathbb{N}$ to be the product measure $\Pi\{p, q\}$, $0 < p, q < 1, p \neq q, p+q = 1$, then the odometer action τ on $\{0,1\}^\mathbb{N}$ is non-singular and ergodic. However no translate of m is symmetric (see J. Aaronson [1]). Note that $\{0,1\}^\mathbb{N}$ as the group of diadic integers has the dual $\{q : q = e^{2\pi \frac{ki}{2^n}}, k \in \mathbb{Z}, n \in \mathbb{N}\}$.

13.10. We now discuss example **3.12**. In this example we have $G = S^1, \tau x = \alpha x$, where $\{\alpha^n\}$, $n \in \mathbb{Z}$, is dense in S^1. Further $m = $ Haar measure on G. Now $A(1, z) = z^p$. The spectral measure F of $V_A = V$ is of Lebesgue type with multiplicity p. Thus $l = p$ (see **13.4**) and C is therefore cohomologous to a $p \times p$ matrix valued cocycle. Indeed, in this case one can calculate C. It is given by the $p \times p$ matrix

$$C(1, z) = \begin{pmatrix} 0 & 0 & \cdots & z^p \\ 1 & 0 & 0 & \cdots & 0 \\ \vdots & & & \vdots & \vdots \\ 0 & 0 & 0 & \cdots & 1 \end{pmatrix}.$$

This example shows that we can get irreducible systems of imprimitivity of multiplicity greater than one. For $p \geq 0$, $A(1, z) = z^p$ is an inner function. It is interesting to note that $C(1, z)$ is also an inner function, although a matricial one. This fact holds more generally in the following sense. If we take (G, τ) as in this example, m the Haar measure on G, and if $V^n, n \in \mathbb{Z}$, acts on $L^2(G, m, \mathcal{H})$ by

$$(Vf)(x) = A(x)f(\tau x), f \in L^2(G, m, \mathcal{H}),$$

where A is a $\mathcal{U}(\mathcal{H})$ valued inner function, then F has the same null sets as the Haar measure on S^1. If $(V^n, n \in \mathbb{Z}, E)$ is irreducible then $C = \hat{A}$ is cohomologous to an inner function (matricial or scalar depending on the multiplicity of F).

The Group of Quasi-Invariance; Its Topology

13.11. Let μ be a finite measure on the circle. We will now show that the group

$$H(\mu) = \{h \in S^1 : \mu_h \text{ and } \mu \text{ have the same null sets}\},$$

is the eigenvalue group of some compact group rotation. We first discuss the unique Polish topology on $H(\mu)$ afresh. For $h \in H(\mu)$ define a unitary operator U_h by

$$U_h f(x) = \left(\frac{d\mu_h}{d\mu}\right)^{1/2} (x)f(hx), \quad f \in L^2(S^1, \mu). \tag{A}$$

We topologise $H(\mu)$ by requiring that $h_n \to h$ if and only if $U_{h_n} f \to U_h f$ for all f. This topology, in fact, is the restriction to $H(\mu)$ of the coarse topology on all non-singular automorphisms of (S^1, μ) (see chapter **8**). We can give a metric to this topology as follows: Choose a complete orthonormal set $\phi_n, n \in \mathbb{N}$, in $L^2(S^1, \mu)$ and put

$$d(p, q) = \sum_{n=1}^{\infty} \frac{1}{2^n}(\|U_p\phi_n - U_q\phi_n\| + \|U_p^{-1}\phi_n - U_q^{-1}\phi_n\|).$$

This topology is separable because the topology of strong convergence on the class of all unitary operators of a complex separable Hilbert space is second countable. We check that d is complete. Let $q_n, n \in \mathbb{N}$, be a Cauchy sequence in this metric. Then $U_{q_n}, n \in \mathbb{N}$, converges in the strong operator topology to a unitary operator U which moreover preserves positive functions (since each U_{q_n} has this property). If we define V by

$$(Vf)(x) = xf(x), f \in L^2(S^1, \mu),$$

then

$$U_{q_n}Vf = q_n V U_{q_n} f, \quad \text{for all } f \in L^2(S^1, \mu).$$

It follows that $q_n, n \in \mathbb{N}$, must converge to an element $q \in S^1$ (in the usual topology of S^1) and $UV = qVU$, or, $UVU^{-1} = qV$. Now V has the spectral resolution

$$V = \int_{S^1} x dF, \text{ where } F(A)f = 1_A f, \quad A \subseteq S^1, \; f \in L^2(S^1, \mu),$$

and $UVU^{-1} = qV$ implies that $UF(A)U^{-1} = F(q^{-1}A)$. This shows that $q \in H(\mu)$ and U has the form

$$(Uf)(x) = C(x) \left(\frac{d\mu_q}{d\mu} \right)^{1/2} (x) f(qx),$$

where C is a function of absolute value one. Since U preserves non-negative functions, $C(x) = 1$ for all x and so $U = U_q$. This proves the completeness of the metric d. Clearly the injection map $i : H(\mu) \to S^1$ is continuous and q_n converges to q in this metric if and only if q_n converges to q in the usual topology and $\frac{d\mu_{q_n}}{d\mu}$ converges to $\frac{d\mu_q}{d\mu}$ in measure (see chapter **9**). Note that if μ is ergodic under $H(\mu)$ then μ is ergodic under any countable dense subgroup of $H(\mu)$ (dense in the new topology of $H(\mu)$).

The Group of Quasi-Invariance; It is an Eigenvalue Group

13.12 Theorem. *Let μ be ergodic with respect to the $H(\mu)$ action on S^1. Then there is a compact group rotation (G, τ) and a finite measure m on G, quasi-invariant and ergodic under τ such that $e(\tau)$, the eigenvalue group of τ (with respect to m), and $H(\mu)$ are the same. Moreover there is a continuous one-one homomorphism of $e(\tau)$ into $E(\tau)$, the group of eigenfunctions of absolute value one.*

Proof. Let $Q \subseteq H(\mu)$ be a countable dense subgroup in the topology of $H(\mu)$. Put $G = \hat{Q}_d$. Let $U_q, q \in Q$, be given by (A). Let

$$U_q = \int_G \chi_q dE, \quad q \in Q,$$

be the spectral resolution of $U_q, q \in Q$, where χ_q denotes the character on G corresponding to $q \in Q$. With V given by

$$(Vf)(x) = xf(x), \; f \in L^2(S^1, \mu),$$

and F its spectral resolution we have

$$V^n U_q = q^{-n} U_q V^n$$

which implies that

$$U_q^{-1} F(A) U_q = F(Aq^{-1}), \quad q \in H(\mu), \quad A \subseteq S^1, \tag{3}$$

$$V^n E(B) V^{-n} = E(B + nx_0) = E(\tau^n B), \quad n \in \mathbb{Z}, \quad B \subseteq G. \tag{4}$$

Here $x_0(q) = q$, and $\tau x = x + x_0$. The pair $(U_q, q \in Q, F)$ is a system of imprimitivity based on (S^1, Q) which is irreducible since F has multiplicity one. This in turn forces the system $(V^{-n}, n \in \mathbb{Z}, E)$ to be irreducible. The spectral measure E therefore has uniform multiplicity, say, $k \leq \aleph_0$. Let m be a finite measure having the same null sets as E. Now (4) implies that m is quasi-invariant under τ and the irreducibility of the system $(V^{-n}, n \in \mathbb{Z}, E)$ implies that τ is ergodic with respect to m. We show that $e(\tau) = H(\mu)$. Let $\mathcal{H}$ be a k dimensional Hilbert space. By the Hahn-Hellinger theorem there exists an isometry $S : L^2(S^1, \mu) \leftrightarrow L^2(G, m, \mathcal{H})$ such that

$$SE(A)S^{-1} f = 1_A f, \ f \in L^2(G, m, \mathcal{H}), \ A \subseteq G,$$

$$SU_q S^{-1} f = \chi_q f, \ f \in L^2(G, m, \mathcal{H}), \ q \in Q.$$

We write $\tilde{U}_q = SU_q S^{-1}$. If $\tilde{V} = SVS^{-1}$ then $\tilde{V}^n, n \in \mathbb{Z}$, acts in the following manner: There is a $\mathbb{Z} \times G$ cocycle A taking values in the group $U(\mathcal{H})$ of all unitary operators on $\mathcal{H}$ such that for all $f \in L^2(G, m, \mathcal{H})$

$$(\tilde{V}^{-n} f)(x) = A(n, x) \left(\frac{dm_{\tau^n}}{dm} \right)^{1/2} (x) f(\tau^n x) \quad \text{a.e. } m.$$

Further, for $h \in H(\mu)$, U_h commutes with all $U_q, q \in Q$, so that $\tilde{U}_h = SU_h S^{-1}$ commutes with all $\tilde{U}_q, \ q \in Q$. This shows that $\tilde{U}_h$ must be of the form

$$(\tilde{U}_h f)(x) = u_h(x) f(x), \quad \text{a.e. } m.$$

where u_h is a measurable function taking values in the space of bounded linear operators in $\mathcal{H}$. Since $\tilde{U}_h$ is unitary u_h has values in $\mathcal{U}(\mathcal{H})$. Since Q is dense in $H(\mu)$, we see that whenever $q_n \to h \in H(\mu)$, $q_n \in Q$, $U_{q_n} \to U_h$ in the strong operator topology, whence $\tilde{U}_{q_n} \to \tilde{U}_h$ in the strong operator topology. This shows that u_h is indeed a scalar function of absolute value one, being the L^2 limit of χ_{q_n}'s. We see that u_h is an eigenfunction with eigenvalue h because

$$u_h(\tau x) = \lim_{n \to \infty} \chi_{q_n}(\tau x) = \lim_{n \to \infty} q_n \chi_{q_n}(x) = h u_h(x)$$

(where the limit is taken over a subsequence over which χ_{q_n} converges a.e.). If λ is another eigenvalue of τ with eigenfunction u_λ of absolute value one, then we can define

$$(\tilde{U}_\lambda f)(x) = u_\lambda(x) f(x), f \in L^2(G, m, \mathcal{H}).$$

Now
$$\tilde{V}\tilde{U}_\lambda = \lambda^{-1}\tilde{U}_\lambda\tilde{V}.$$

If $U_\lambda = S^{-1}\tilde{U}_\lambda S$, then $VU_\lambda = \lambda^{-1}U_\lambda V$, which implies that for all Borel $A \subseteq S^1$,
$$U_\lambda^{-1}F(A)U_\lambda = F(A\lambda^{-1}),$$

so that $\lambda \in H(\mu)$. This shows that $H(\mu) = e(\tau)$. Finally the mapping $h \to u_h$ is indeed a continuous one-one homomorphism of $e(\tau)$ into the group of eigenfunctions of absolute value one (with the L^2 topology). The theorem is proved.

13.13. If μ of theorem **13.12** is singular to Haar measure then m cannot be discrete and $H(\mu) = e(\tau)$ is a saturated subgroup of S^1 (see Chapter 14). Moreover it is a proper subgroup, hence a weak Dirichlet set.

Extensions of Cocycles

13.14. We return to the notation of **13.4** and consider an irreducible system described by the pair (C, μ) where C is a $Q \times S^1$ cocycle taking values in $\mathcal{U}(\mathcal{H})$, where $\mathcal{H}$ is a complex separable Hilbert space of dimension k and μ is quasi-invariant and ergodic under Q. Let $(A, m) = (C, \mu)^\wedge$. Let $\lambda \in H(\mu) - Q$.

13.15 Definition. We say that C has an extension to λ if C has an extension to the group $Q_1 = Q(\lambda)$ generated by Q and λ, i.e., there exists a $Q_1 \times S^1$ cocycle C_1 which agrees with C on $Q \times S^1$.

13.16 The question of extension of cocycles is discussed in Helson [4] in the special case when μ is the Haar measure and C is scalar valued. It is shown there that an extension need not always exist (corollary **13.23**), and that the notion of extension is related to a problem in Diophantine approximation.

13.17. We will state a necessary and sufficient condition for an extension of C to exist and derive some consequences of it.

13.18. A complex number λ is said to be an (A, m) eigenvalue of τ if there exists a $\mathcal{U}(\mathcal{K})$ valued measurable function B on G such that
$$A(x)B(\tau x) = \lambda B(x)A(x) \ a.e. \ m.$$

The function B is called an (A, m) eigenfunction with eigenvalue λ. An (A, m) eigenvalue λ is said to be compatible with Q if there is an (A, m) eigenfunction B_λ such that whenever $\lambda^n \in Q$ we have $B_\lambda^n = \chi_{\lambda^n} I$.

13.19 Theorem. *The cocycle C has an extension to $\lambda \in H(\mu) - Q$ if and only if λ is an (A, m) eigenvalue compatible with Q.*

We omit the proof of this theorem and refer the reader to J. Aaronson and M. Nadkarni [2]. The method of proof is much the same as that of theorem **13.12**. We give below some corollaries of this theorem.

13.20 Corollary. *If λ is an L^∞ eigenvalue of τ with respect to the measure m then the cocycle C has an extension to λ.*

Proof. If λ is an L^∞ eigenvalue of τ with respect to m, then $\phi_\lambda I$ is an (A, m) eigenfunction compatible with Q, where ϕ_λ is an eigenfunction of absolute value one corresponding to λ.

13.21 Corollary. *If E has multiplicity one then C extends to λ if and only if λ is an L^∞ eigenvalue of τ with respect to m.*

13.22 Corollary. *If E has multiplicity one and m has the same null sets as the Haar measure on G, then C has no extension to any $\lambda \in H(\mu) - Q$.*

Proof. The L^∞ eigenvalue group of τ with respect to the Haar measure on G is the dual group Q. Hence C cannot have an extension to any λ not in Q.

13.23 Corollary. *If E has finite multiplicity, say k, and has the same null sets as the Haar measure on G, then C has no extension to any λ which satisfies $\lambda^k \notin Q$.*

Proof. If C has an extension to such a λ, λ^k is an L^∞ eigenvalue of τ with respect to Haar measure which is a contradiction.

Suppose now that $C = I$ and the dimension of K is one. Then we have

$$(U_q f)(x) = \left(\frac{d\mu_q}{d\mu}(x) \right)^{1/2} f(qx), \quad q \in Q, \quad f \in L^2(S^1, \mu).$$

The cocycle C has an extension to all of $H(\mu)$ since $C = I$. We have:

13.24 Theorem. *If $C = I$ and the dimension of H is one, then $U_q, q \in Q$, does not have Haar spectrum of finite multiplicity whenever there is a $\lambda \in H(\mu) - Q$ such that for all n, $\lambda^n \notin Q$. In particular if $H(\mu)$ is uncountable, $U_q, q \in Q$, does not have Haar spectrum of finite multiplicity.*

13.25. It is not known if $U_q, q \in Q$, of **13.24** can have Haar spectrum except when μ is discrete.

13.26. Some Open Questions. Let τ be a non-singular automorphism on a probability space $(X, \mathcal{B}, m)$. For each $t \in e(\tau)$ let χ_t be an eigenfunction of absolute value one. The unitary operators

$$(V^n f)(x) = \left(\frac{dm_{\tau^n}}{dm} \right)^{1/2} (x) f(\tau^n x), \ f \in L^2(X, \mathcal{B}, m),$$

$$(U_q f)(x) = \chi_q(x) f(x), \ f \in L^2(X, \mathcal{B}, m), \ q \in Q,$$

satisfy the commutativity relation of Weyl and von Neumann $(V^n U_q = q^n U_q V^n)$, from which we conclude that the spectral measure F of $V^n, n \in \mathbb{Z}$, satisfies, with $U_q, q \in e(\tau)$, the identity $U_q^{-1} F(A) U_q = F(qA)$, $A \subseteq S^1$ Borel. The maximal spectral type μ of F is therefore quasi-invariant under $e(\tau)$. In case the eigenfunctions $\chi_q, q \in Q$, generate $\mathcal{B}$ (mod m), μ is ergodic under $e(\tau)$ and the system of imprimitivity $(U_q, q \in Q, F)$ is irreducible. It is not known if the converse holds, i.e., whether ergodicity of μ under $e(\tau)$ or the irreducibility of $(U_q, q \in Q, F)$ implies that $\chi_q, \ q \in e(\tau)$, generates the σ-algebra $\mathcal{B}$. Assume that $\chi_q, q \in e(\tau)$, generates $\mathcal{B}$ (mod m). In this case we can generate factors of τ as follows: Let $\mathcal{B}_0$ be a σ-algebra generated by a subcollection of eigenfunctions. Then $\mathcal{B}_0$ can easily be verified to be invariant under τ, and hence a factor of τ. It is not known if every factor of τ is of this form, the assumption that $\chi_q, q \in Q$, generate $\mathcal{B}$ still being in force. It is not even known in this case if every factor must have eigenvalues. Under the same assumption it is not known whether $e(\tau) = H(\mu)$ always holds, where μ is the maximal spectral type of F. If the answer to this question is affirmative, then for a compact group rotation τ, the unitary operator V has Lebesgue spectrum if and only if m is discrete.

23.27. Our exposition in this chapter has relied mainly on J. Aaronson and M. Nadkarni [2].

Chapter 14

Saturated Subgroups of the Circle Group

Saturated Subgroups of S^1

14.1. Consider the embedding of $\mathbb{Z}$ into $L^1(S^1, \mu)$, μ a probability measure, given by $n \to z^n$. The collection $\{z^n : n \in \mathbb{Z}\}$ is discrete in $L^1(X, \mathcal{B}, \mu)$ if and only if

$$\limsup_{n \to \infty} |\hat{\mu}(n)| < 1.$$

This result, due to C. C. Moore and K. Schmidt [3], shows that such a measure μ is non-rigid, hence not supported on a Dirichlet set. Such measures may be viewed as being full in some sense even in the case where μ is singular. More generally, given two probability measures μ and ν on S^1, one can map, for each n, z^n in $L^1(S^1, \mu)$ to z^n in $L^1(S^1, \nu)$, and seek conditions under which this map extends to a continuous homomorphism between the closures of characters in the respective spaces. We will answer this question in this chapter and discuss its relation to subgroups of the circle group such as the eigenvalue group or the group of quasi-invariance of a measure. As we saw in the previous chapters, such subgroups occur naturally in non-singular dynamics.

14.2. A measurable subgroup $H \subseteq S^1$ is said to be saturated if for any $\mu \in M(S^1)$,

$$|\mu(H)| \leq \sup_{n \in \mathbb{Z}} |\hat{\mu}(n)|,$$

where $M(S^1)$ stands for the collection of all finite complex measures on S^1.

14.3 Theorem. *For a measurable subgroup $H \subseteq S^1$, the following are equivalent:*

(1) *H is saturated, equivalently, $|\mu(H)| \leq \sup_{n \in \mathbb{Z}} |\hat{\mu}(n)|$,*

(2) *$\forall$ compact $K \subseteq H$, $\forall$ compact L disjoint from H, $\forall \varepsilon > 0$, $\exists$ a positive definite continuous function ϕ with $\phi(1) = 1$ such that*

$$|1 - \phi(t)| \leq \varepsilon, \quad t \in K,$$
$$|\phi(t)| \leq \varepsilon, \quad t \in L,$$

(3) *for every positive finite measure* μ, 1_H *is in the closed convex hull in* $L^1(S^1, \mu)$ *of continuous characters* χ_n, $n \in \mathbb{Z}$.

Proof. Recall that a continuous positive definite function on S^1 is the Fourier transform of a finite non-negative measure on $\mathbb{Z}$, hence the sum of an absolutely convergent Fourier series with non-negative Fourier coefficients. The condition $\phi(1) = 1$ means that the sum of the Fourier coefficients of ϕ is one. Thus ϕ belongs to the closed convex hull of continuous characters under the uniform norm.

(1) $\Rightarrow$ **(2)**. Consider the space of real continuous function on $K \cup L$ with uniform norm. It is enough to prove that 1_K is in the closed convex hull of the functions $\Re\chi_n$, $n = 0, 1, 2, \dots$. If 1_K is not in the closed convex hull of $\Re\chi_n$, $n = 0, 1, 2, \dots$, then by the separating hyperplane theorem there is a real measure λ supported on $K \cup L$ and a constant a such that for all $n \in \mathbb{Z}$,

$$\Re\hat{\lambda}(n) = \int_{S^1} (\Re\chi_n) d\lambda < a < \lambda(H) = \int_{S^1} 1_K d\lambda.$$

Let μ be a symmetric measure $k\delta + \frac{1}{2}(\lambda + \tilde{\lambda})$, where δ denotes the Dirac mass at 1, and k is such that for all n, $\hat{\mu}(n) = k + \Re\hat{\lambda}(n) > 0$, $\mu(H) = k + \lambda(H) > 0$. We then have for all $n \in \mathbb{Z}$,

$$\mu(H) = k + \lambda(H) > k + a > k + \Re\hat{\lambda}(n) = \hat{\mu}(n),$$

which is a contradiction since H is assumed to be saturated.

(2) $\Rightarrow$ **(3)**. Given any probability measure μ on S^1 and $\varepsilon > 0$ find a compact set $K \subseteq H$ and a compact set L disjoint from H such that $\mu(K \cup L) > 1 - \varepsilon$. Since the function ϕ given in **(2)** is a uniform limit of convex combinations of characters, **(3)** is a direct consequence of **(2)**.

(3) $\Rightarrow$ **(1)**. Let μ be a complex measure on S^1. By property **(3)** there exists a sequence (f_n), $n \in \mathbb{N}$, of convex combinations of characters such that $f_n \to 1_H$ in $L^1(S^1, |\mu|)$. Clearly

$$|\mu(H)| = \left| \int_{S^1} 1_H d\mu \right| = \lim_{n \to \infty} \left| \int_{S^1} f_n d\mu \right| \leq \sup_{n \in \mathbb{Z}} \left| \int_{S^1} \chi_n d\mu \right|,$$

so that **(3)** $\Rightarrow$ **(1)**, and the theorem is proved.

14.4 Corollary. *A measurable subgroup* $H \subseteq S^1$ *is saturated if and only if for some constant* $c > 0$,

$$|\mu(H)| \leq c \sup_{n \in \mathbb{Z}} |\tilde{\mu}(n)|, \ \forall\, \mu \in M(S^1).$$

Proof. If H is saturated then the conclusion of the corollary is satisfied with $c = 1$. On the other hand if $|\mu(H)| \leq c \sup_{n \in \mathbb{Z}} |\hat{\mu}(n)|$ for all μ in $M(S^1)$ for some c and if H is not saturated, we arrive at a contradiction as follows: Pick a measure μ in $M(S^1)$ such that $\mid \mu(H) \mid \geq 1$ and $\sup_{n \in \mathbb{Z}} \mid \hat{\mu}(n) \mid \leq 1 - \varepsilon$, for some $\varepsilon > 0$. Consider the measure $\nu = \mu * \tilde{\mu}$, where $\tilde{\mu}(A) = \overline{\mu}(A^{-1})$, $A \subset S^1$. We have

$$\nu(H) = \int_{S^1} \mu(H + x)d\overline{\mu}.$$

Now there exist at most countably many disjoint classes $H + x$ with $\mu(H + x) \neq 0$; say $\mu(H + x) = 0$ if x does not belong to one of the classes $H + x_n$, $n \in \mathbb{N}$, so that

$$\begin{aligned}
\nu(H) &= \sum_{n \geq 1} \int_{H + x_n} \mu(H + x)d\overline{\mu} = \sum_{n \geq 1} \mu(H + x_n)\overline{\mu}(H + x_n) \\
&= \sum_{n \geq 1} \mid \mu(H + x_n) \mid^2 \; \geq 1.
\end{aligned}$$

On the other hand $\hat{\nu}(n) = \mid \hat{\mu}(n) \mid^2$ and $\sup_{n \in \mathbb{Z}} \mid \hat{\nu}(n) \mid \leq (1 - \varepsilon)^2$. By iterating this argument we will get for all $k \geq 1$, and $n \in \mathbb{Z}$, $\nu^k(H) \geq 1$, $\mid \hat{\nu}^k(n) \mid \leq (1 - \varepsilon)^k$. Clearly for large enough k the hypothesis of the corollary will be contradicted by ν^k.

14.5. For any measurable subgroup H of the circle, the function 1_H is positive definite and Borel. Property **(3)** of theorem **14.3** asserts that H is saturated if and only if given any positive measure μ, 1_H is the limit in $L^1(S^1, \mu)$ (or μ a.e.) of a sequence of continuous positive definite functions. Note that S^1 is a saturated subgroup since for any μ in $M(S^1)$, $\mid \mu(S^1) \mid = \mid \hat{\mu}(0) \mid$.

14.6 Corollary. *A proper subgroup H of S^1 is saturated if and only if for all $\mu \in M(S^1)$.*

$$\mid \mu(H) \mid \leq \limsup_{n \to \infty} \mid \hat{\mu}(n) \mid .$$

Proof. Clearly if the condition above holds then H is a saturated subgroup. On the other hand suppose H is a saturated proper subgroup of S^1. Since H is a proper subgroup its Lebesgue measure is zero. (Otherwise $H = H - H$ has non-empty interior and $H = S^1$). Let $\mu \in M(S^1)$. Given any positive integer N, we can find an absolutely continuous measure ν (with respect to the Lebesgue measure) such that $\hat{\nu}(n) = \hat{\mu}(n)$ for $-N \leq n \leq N$, $\hat{\nu}(n) = 0$ otherwise. Then $\nu(H) = 0$ and

$$\mid \mu(H) \mid = \mid (\mu - \nu)(H) \mid \leq \sup_{n \in \mathbb{Z}} \mid (\hat{\mu} - \hat{\nu})(n) \mid = \sup_{|n| \geq N} \mid \hat{\mu}(n) \mid .$$

Since N is arbitrary the result is established.

Relation to Closures and Convex Hulls of Characters

14.7. We would like to give some more characterisations of saturated subgroups. To this end we introduce, for a positive measure μ on S^1, the objects:

(1) $\overline{Z}_1(\mu)=$ closure of continuous characters in $L^1(S^1,\mu)$,
(2) $\overline{Z}(\mu) =$ closure of continuous characters in the weak* topology of $L^\infty(S^1,\mu)$,
(3) $\tilde{Z}(\mu) =$ closed convex hull of continuous characters in the weak* topology of $L^\infty(S^1,\mu)$, also equal to the closed convex hull of continuous characters in $L^1(S^1,\mu)$ topology.

14.8. $\overline{Z}_1(\mu)$ is a closed subgroup of functions of unit modulus in $L^1(S^1,\mu)$. We may regard $\mathbb{Z}$ as embedded in the group of functions of unit modulus in $L^1(S^1,\mu)$ via the mapping $n \to \chi_n$, and, $\overline{Z}_1(\mu)$ may be viewed as the completion of the integers with respect to the invariant metric

$$d(m,n) = \int_{S^1} |\chi_m - \chi_n|\, d\mu.$$

It is easy to see that the completion of continuous characters in the $L^p(S^1,\mu)$ topology for any p, $1 \le p < \infty$, is the same as $\overline{Z}_1(\mu)$. It is also true that for any sequence (n_k) of integers, $\chi_{n_k} \to 1$ in $L^1(S^1,\mu)$ if and only if $\chi_{n_k} \to 1$ in the weak* topology of $L^\infty(S^1,\mu)$, and this is equivalent to saying that $\hat{\mu}(n_k) \to \|\mu\|$ as $k \to \infty$.

14.9. The sets $\overline{Z}(\mu)$ and $\tilde{Z}(\mu)$ form semigroups under pointwise multiplication and the operation is separately continuous in each variable; $\overline{Z}(\mu)$ and $\tilde{Z}(\mu)$ are compact; both contain $\overline{Z}_1(\mu)$.

14.10 Lemma. *For positive finite measures μ and ν on S^1 the following are equivalent:*

(1) *For any sequence n_k, $k \in \mathbb{N}$, $\lim_{k\to\infty} \hat{\mu}(n_k) = \|\mu\|$ implies that*

$$\lim_{k\to\infty} \hat{\nu}(n_k) = \|\nu\|.$$

(2) *For any sequence n_k, $k \in \mathbb{N}$,*

$$\lim_{k\to\infty} \int_{S^1} |\chi_{n_k} - 1|\, d\mu = 0 \Rightarrow \lim_{k\to\infty} \int_{S^1} |\chi_{n_k} - 1|\, d\nu = 0.$$

(3) *Every element of $\overline{Z}(\mu+\nu)$ which is 1 a.e. μ is 1 a.e. $\mu+\nu$.*
(4) *There exist a continuous group homomorphism from $\overline{Z}_1(\mu)$ to $\overline{Z}_1(\nu)$ which maps, for each $l \in \mathbb{Z}$, the function χ_l of $\overline{Z}_1(\mu)$ to the function χ_l of $\overline{Z}_1(\nu)$.*
(5) *Every element of $\tilde{Z}(\mu+\nu)$ which is 1 a.e. μ is also 1 a.e $\mu+\nu$.*

Proof. Equivalence of **(1)**,**(2)**, **(3)**, **(4)** follows immediately from remarks in **14.8** and **(5)** obviously implies **(3)**. It remains to prove that **(3)** implies **(5)**.

Every element ψ of $\tilde{Z}(\mu + \nu)$ can be written as a barycenter of elements in $\overline{Z}(\mu+\nu)$, say $\psi = \int u d\sigma(u)$ with some probability measure σ on $\overline{Z}(\mu+\nu)$. Now if $\psi = 1$ μ a.e. then necessarily for σ-almost every u, we must have $u = 1$ μ a.e., whence by (3) $u = 1$ $(\mu + \nu)$ a.e., and finally, $\psi = 1$ $(\mu + \nu)$ a.e.

14.11 Definition. Let μ and ν be positive measures on S^1. If the equivalent properties (1) to (5) of the theorem above hold with respect μ and ν then we say that ν sticks to μ. More generally, we say that a complex measure ν sticks to a complex measure μ if $| \nu |$ sticks to $| \mu |$.

14.12. If for a positive measure μ, $\limsup_{n\to\infty} | \hat{\mu}(n) | < \| \mu \|$, then every measure ν sticks to μ. We see this as follows: For such a μ the requirement $\lim_{k\to\infty} \hat{\mu}(n_k) \to \| \mu \|$ implies that (n_k), $k \in \mathbb{N}$, is a bounded sequence and so a certain integer m is repeated infinitely often in the sequence, and $| \hat{\mu}(m) | = \| \mu \|$. Clearly $\chi_m = 1$ a.e.μ so that $\hat{\mu}(mn) = \| \mu \|$ for all n. We must have $m = 0$ so that $n_k = 0$ from a certain stage onwards. Clearly then $\hat{\nu}(n_k) = \| \nu \|$ for all large k, and ν sticks to μ.

14.13 Proposition. *Let μ be a positive finite measure on S^1. The map ψ : $n \to \chi_n$ of $\mathbb{Z}$ into the group of functions of unit modulus in $L^1(S^1, \mu)$ is a homeomorphism if and only if $\limsup_{n\to\infty} | \hat{\mu}(n) | < \| \mu \|$.*

Proof. It is enough to show that the characters $\chi_n, n \in \mathbb{Z}$, form a discrete set in $L^1(S^1, \mu)$ if and only if $\limsup_{n\to\infty} | \hat{\mu}(n) | < \| \mu \|$. Now the set of continuous characters is not discrete in $L^1(S^1, \mu)$ if and only if given $\varepsilon > 0$ there exist distinct continuous characters χ_m, χ_n such that $\| \chi_m - \chi_n \|_1 < \varepsilon$, which holds if and only if there exists a sequence of continuous characters converging to 1 a.e. μ, i.e., if and only if $\limsup_{n\to\infty} | \hat{\mu}(n) | = \| \mu \|$, contrary to the assumption.

14.14 Theorem. *A measurable subgroup H of S^1 is saturated if and only if every measure which sticks to a measure concentrated on H is itself concentrated on H.*

Proof. Recall (3) of theorem **14.3** which we can state in the form: H is saturated if and only if for any positive measure $\mu \in M(S^1)$, $1_H \in \tilde{Z}(\mu)$. Assume that H is a saturated subgroup of S^1 and let μ and ν be two positive measures such that μ is concentrated on H and ν sticks to μ. Since 1_H belongs to $\tilde{Z}(\mu+\nu)$, property (5) of lemma **14.10** yields that $1_H = 1$ a.e. ν and therefore ν is concentrated on H.

Conversely, let μ be any positive measure on S^1. The set of non-negative elements of $\tilde{Z}(\mu)$ which are $\geq 1_H$ a.e. μ is compact in the weak* topology and thus admits a minimal element h for the natural order relation on the μ-measurable functions. Since $\tilde{Z}(\mu)$ is a multiplicative semigroup, h is a zero-one function. We claim that the measure $\nu = h\mu$ sticks to the measure $\mu_H = 1_H\mu$. Let ϕ be any element of $\tilde{Z}(\nu)$ which is equal to 1 a.e. μ_H. Then ϕ is a limit in the weak* topology on $L^\infty(S^1, \nu)$ of a sequence of convex combinations of continuous characters. Members of this sequence may be viewed as belonging to the unit ball of $L^\infty(S^1, \mu)$, hence will have a weak* limits, one of which we denote by ϕ'. This ϕ' will extend ϕ to an element in $\tilde{Z}(\mu)$, $\phi' = 1$ a.e. μ_H. Then $\psi = \frac{1}{2}(1 + \Re(\phi'))$ is a positive element of $\tilde{Z}(\mu)$ with $\psi = 1_H$ a.e. μ. Thus $\psi \geq h$, so that $\psi = 1$ a.e. ν. It follows that $\phi = \phi' = 1$ a.e. ν. By property (5) of lemma 14.10, ν sticks to μ_H. Finally, under the conditions of the theorem, ν is concentrated on H; this proves that $h = 1_H$ a.e. μ, so that $1_H \in \tilde{Z}(\mu)$. This verifies (3) of theorem 14.3 and H is saturated.

14.15 Proposition. *Any countable subgroup H of S^1 is saturated.*

Proof. We give H the discrete topology and let $\hat{H}$ denote the compact dual of H. We can embed $\mathbb{Z}$ in $\hat{H}$ by $n \to \chi_n$, where $\chi_n(u) = u^n$, $u \in S^1$. This embedding of $\mathbb{Z}$ in $\hat{H}$ is dense in $\hat{H}$. Let μ and ν be two positive measures such that μ is concentrated on H and ν sticks to μ. If χ_{n_k}, $k \in \mathbb{N}$, converges pointwise on H, it converges in $\overline{Z}_1(\mu)$, and by property (4) of lemma 14.10, it converges in $\overline{Z}_1(\nu)$, so that $\hat{\nu}(n_k)$, $k \in \mathbb{N}$, converges. Therefore the Fourier transform $\hat{\nu}$ (viewed as a function defined on the embedding of $\mathbb{Z}$ in $\hat{H}$) may be extended to a continuous function on $\hat{H}$. It follows from Bochner's theorem that ν is concentrated on H. The proposition is proved.

14.16. More generally, given a measurable subgroup H, we can define a group topology on $\mathbb{Z}$ such that a sequence n_k, $k \in \mathbb{N}$, converges to zero if and only if for every positive measure μ carried by H, the sequence $\hat{\mu}(n_k) \to \| \mu \|$ as $k \to \infty$. By **14.13** H is saturated when any positive measure whose Fourier transform is continuous in this topology is concentrated on H.

σ-Compact Saturated Subgroups; H_2 Groups

14.17. The next theorem is a more precise formulation of property (2) of theorem **14.3** for the case when H is σ-compact. It permits us to provide non-trivial examples of saturated subgroups.

14.18 Theorem. *Let H be a σ-compact subgroup of S^1. The following are equivalent:*

(1) *H is saturated,*
(2) *For any compact L disjoint from H, one can find a sequence $\phi_j, j = 0, 1, 2, \ldots$, of real positive definite functions with $\phi_j(1) = 1$ for all j and such that*

$$\sum_{j=0}^{\infty} | 1 - \phi_j(t) |$$

is finite for all $t \in H$ and infinite for all $t \in L$.

(3) *For any compact L disjoint from H, one can find a sequence $a_j, j = 0, 1, 2, \ldots$, of non-negative real numbers, and a sequence $n_j, j = 0, 1, 2, \ldots$, of positive integers (not necessarily distinct) such that the series*

$$\sum_{j=0}^{\infty} a_j (1 - \Re \chi_{n_j}(t))$$

is finite for all $t \in H$ and infinite for all $t \in L$.

Proof. **(1)** $\Rightarrow$ **(2)**. Since H is σ-compact, it can be written as an increasing union of sequences of compact $K_j, j \geq 0$. Since H is saturated, by **(2)** of **14.3** we can find for each K_j a continuous positive definite function ψ_j with $\psi_j(1) = 1$

$$|1 - \psi_j| < \frac{1}{2^j} \quad \text{on} \quad K_j, \quad |\psi_j| < \frac{1}{2^j} \quad \text{on} \quad L.$$

If $\phi_j = \Re(\psi_j)$, then $\phi_j(1) = 1$, ϕ_j is continuous, positive definite, and $\sum_{j=0}^{\infty}(1 - \phi_j(t))$ is finite for all $t \in H$ and infinite for all $t \in L$.

(2) $\Rightarrow$ **(3)** For each $j \geq 0$ the function ϕ_j provided by **(2)** is real continuous and positive definite, hence it is the Fourier transform of a symmetric probability measure on $\mathbb{Z}$:

$$\phi_j(t) = \sum_{k=-\infty}^{\infty} a_{j,k} \chi_{-k}(t), \quad a_{j,k} = a_{j,-k} \geq 0, \quad \sum_{k=-\infty}^{\infty} a_{j,k} = 1.$$

Now

$$\sum_{j=0}^{\infty}(1 - \phi_j(t)) = \sum_{j=0}^{\infty} \sum_{k=1}^{\infty} 2a_{j,k}(1 - \Re \chi_k(t)),$$

which is a series of the type mentioned in **(3)**. (Note that the integers n_j in the series in **(3)** are not required to be distinct). Since the left hand side is finite for $t \in H$ and infinite for $t \in L$, the right hand side has the same property.

(3) $\Rightarrow$ **(1)**. We may assume that $a_j \leq 1$ for all j (by splitting when needed). We shall show that property **(3)** of theorem **14.3** holds. Given any positive measure μ and $\varepsilon > 0$, we can find a compact L, disjoint from H, such that $\mu(H \cup L) \geq (1 - \varepsilon) \| \mu \|$. Choose a series $\sum_{j=0}^{\infty} a_j (1 - \Re \chi_{n_j}(t))$ provided by **(2)** for this L. For every $k \geq 1$ we define

$$\psi_k(t) = \prod_{j=0}^{\infty} \left(1 - \frac{1}{k}(a_j - a_j \Re \chi_{n_j}(t)) \right).$$

The ψ_k's are pointwise limits of a sequences of positive definite continuous functions with $\psi_k(0) = 1$, and therefore belongs to $\tilde{Z}(\mu)$. Now ψ_k, as $k \to \infty$, converges to a function ψ given by

$$\psi(t) = 1, \quad \text{if} \quad \sum_{j=0}^{\infty} a_j (1 - \Re \chi_{n_j}(t)) < \infty, \quad = 0 \quad \text{otherwise.}$$

so that $\psi(t) = 1_H(t)$ for all $t \in H \cup L$. We have

$$\psi \in \tilde{Z}(\mu) \quad \text{and} \quad \int_{S^1} |\,\psi - 1_H\,|\,d\mu = \int_{(H \cup L)^c} |\,\psi - 1_H\,|\,d\mu \leq \varepsilon \,\|\,\mu\,\| \,.$$

Thus 1_H is in the closed convex hull in $L^1(S^1, \mu)$ of continuous characters. By condition **(3)** of **14.3**, H is saturated.

14.19 Definition. Let α be a positive real number. Given a sequence n_j, $j \in \mathbb{N}$, of positive integers and a sequence a_j, $j \in \mathbb{N}$, of non-negative real numbers, the set of $t \in S^1$ such that

$$\sum_{j=1}^{\infty} a_j \,|\, 1 - \chi_{n_j}(t) \,|^\alpha < \infty$$

is a subgroup of S^1. A subgroup of S^1 which can be described in this way for some sequences $n_j, j \in \mathbb{N}$, a_j, $j \in \mathbb{N}$, is called an H_α group.

14.20. The set H_α is indeed a group. We can verify this as follows: Clearly $1 \in H_\alpha$ and $t^{-1} \in H_\alpha$ whenever $t \in H_\alpha$. Next

$$1 - \chi_k(tu) = 1 - \chi_k(t) + \chi_k(t) - \chi_k(t)\chi_k(u)$$

$$|\,1 - \chi_k(tu)\,| \;\; \leq \;\; |\,1 - \chi_k(t)\,| + |\,1 - \chi_k(u)\,|$$

$$\leq 2\max\big(\,|\,1 - \chi_k(t)\,|, \quad |\,1 - \chi_k(u)\,|\,\big).$$

If t, u are in H_α, then

$$\sum_{j=0}^{\infty} a_j \,|\, 1 - \chi_{n_j}(tu) \,|^\alpha \leq \sum_{j=0}^{\infty} a_j 2^\alpha \big(\max(|\,1 - \chi_{n_j}(t)\,|, |\,1 - \chi_{n_j}(u)\,|)\big)^\alpha,$$

which is finite. Thus H_α is a group.

The H_2 group corresponding to given sequences a_j, $j \in \mathbb{N}$, and n_j, $j \in \mathbb{N}$, can be described as the set of all $t \in S^1$ such that

$$\sum_{j=0}^{\infty} a_j \,|\, 1 - \Re \chi_{n_j}(t) \,| \;\; < \infty,$$

since for $t \in S^1$, $|\,1 - t\,|^2 = 2 - 2\Re(t) = 2(1 - \Re(t))$ so that

$$\sum_{j=1}^{\infty} a_j \,|\, 1 - \chi_{n_j}(t) \,|^2 < \infty \Leftrightarrow \sum_{j=1}^{\infty} a_j \,|\, 1 - \Re \chi_{n_j}(t) \,| < \infty.$$

14.21 Proposition.

  (i) *Every H_2 group is a saturated σ-compact subgroup of S^1.*

 (ii) *Given any σ-compact saturated subgroup H and a compact L disjoint from H, there exists an H_2 group containing H and disjoint from L.*

Proof. Suppose $H = \{t : \sum_{j=1}^{\infty} a_j \mid 1 - \Re\chi_{n_j}(t) \mid < \infty\}$ is an H_2 group. Then H is σ-compact because the set $\{t : \sum_{j=1}^{\infty} a_j \mid 1 - \Re\chi_{n_j}(t) \mid \leq c\}$ is compact for every $c > 0$. Further, for every $t \notin H$,

$$\sum_{j=1}^{\infty} a_j \mid 1 - \Re\chi_{n_j}(t) \mid = \infty,$$

whence by **(2)** of **14.18** H_2 is saturated.

(ii) This also follows from **(2)** of **14.18** and the definition of H_2 group.

14.22. It is easy to see from **14.18** that if $\phi_j, j \in \mathbb{N}$, is a sequence of real valued continuous positive definite functions with $\phi_j(1) = 1$, then the set of all $t \in S^1$ such that $\sum_{j=1}^{\infty} \mid 1 - \phi_j(t) \mid < \infty$ is an H_2 group and so is a saturated subgroup.

14.23 Proposition. *If $0 < \alpha \leq 2$, then every H_α group is an H_2 group.*

We refer the reader to B. Host, J.-F. Méla, F. Parreau [1] for a proof of this and for the contents of this chapter. Also see K. Schmidt [4], J.-F. Méla [2].

14.24 Remark. Given a sequence $n_j, j \in \mathbb{N}$, with infinitely many non-zero terms, there exists a t_0 such that $\chi_{n_j}(t_0)$ does not converge to 1. If $a_j, j \in \mathbb{N}$, is a sequence of positive real numbers bounded away from zero, then $\sum_{j=1}^{\infty} a_j \mid 1 - \chi_{n_j}(t_0) \mid^2$ does not converge. The set of t for which $\sum_{j=1}^{\infty} a_j \mid 1 - \chi_{n_j}(t) \mid^2$ is finite is thus a proper, σ-compact, saturated subgroup of S^1. More generally, it can be shown that if $\sum_{j=1}^{\infty} a_j = \infty$, then for every sequence $n_j, j \in \mathbb{N}$, with infinitely many non-zero terms the corresponding H_2 group with co-efficients a_j is a proper subgroup of S^1.

14.25 Remark. Every proper H_2 group is contained in a group of the form $D(n_1, n_2, n_3, \ldots) = \{t : \chi_{n_j}(t) \to 1, \text{ as } k \to \infty\}$. Hence every proper H_2 group is a weak Dirichlet set (see **6.13**).

14.26 Remark. It may be surmised that every subgroup of the form $D(n_1, n_2, n_3, \ldots)$ for $n_1 < n_2 < n_3 < \ldots$, is a saturated subgroup. This however is false. Also for $\alpha > 2$, there exist H_α groups which are not saturated [1].

Chapter 15

Riesz Products As Spectral Measures

15.1. In this chapter we will discuss the spectral theory of rank one automorphisms. This is intimately related to the notion of Riesz product. We will define a class of measures on S^1 called Riesz product and show that such measures appear as maximal spectral types of certain towers over adding machine. We will also define generalised Riesz products and show that such measures are related to the spectrum of general rank one automorphisms.

Riesz products were discovered in 1918 (see F. Riesz [25]) to answer affirmatively a special question in the theory of Fourier series, viz., whether there exists a continuous function of bounded variation whose Fourier coefficients are not $o(\frac{1}{n})$, and they were subsequently used to answer other similar questions. Rank one automorphisms appeared around 1965 in connection with some special problems in mathematical ergodic theory. The notion of Riesz product turns out to be exactly the right tool to describe their spectra.

That Riesz products appear as spectral measures of automorphisms was shown by F. Ledrappier [20]. In the exposition below we will adapt the more general formulation of B. Host, J.-F. Méla, F. Parreau [14]. Some recent contributions will be mentioned.

15.2. A word about notation. It is customary to write a trigonometric series in the form $\sum_{n=0}^{\infty}(c_n \cos nx + d_n \sin nx)$ or $\sum_{n=-\infty}^{\infty} a_n e^{inx}$. We shall write $e^{ix} = t$ or z and view our trigonometric polynomials or trigonometric series as being defined on the circle group S^1, rather than on the interval $[0, 2\pi]$. Lebesgue measure and Haar measure on S^1 will mean the same thing and will be denoted by dt or dz.

Dissociated Trigonometric Polynomials

15.3. Consider the product $(1+t)(1+t) = 1 + t + t + t^2$ which, by writing the middle two terms as $2t$, can be written as $1 + 2t + t^2$. On the other hand if we expand the product $(1+t)(1+t^2)$ we get $1 + t + t^2 + t^3$ where we cannot group terms to reduce the number of terms in the polynomial. In the second case we say that the polynomials $(1+t)$ and $(1+t^2)$ are dissociated. More generally:

15.4 Definition. Two trigonometric polynomials

$$p_1(t) = \sum_{k=-M}^{M} c_k t^k, \quad p_2(t) = \sum_{j=-N}^{N} d_j t^j, t \in S^1,$$

are said to be dissociated if, when we consider the formal expansion of their product:

$$p_1(t)p_2(t) = \sum_{k=-M}^{M} \sum_{j=-N}^{N} c_k d_j t^{k+j},$$

the powers $k + j$ of t in the non-zero terms $c_k d_j t^{k+j}$ are all distinct.

15.5 Definition. A finite set $p_0, p_1, p_2, \ldots, p_k$ of trigonometric polynomials $p_j(t) = \sum_{i=-N_j}^{N_j} c_i(j)t^i, j = 0, 1, 2, \ldots, k$ are said to be dissociated if in their product $p_0(t)p_1(t)p_2(t)\cdots p_k(t)$, (when expanded formally, i.e., without grouping terms or cancelling identical terms with opposite signs), the powers $i_0 + i_1 + i_2 + \cdots + i_k$ of t in non-zero terms $c_{i_0}(0)c_{i_1}(1)c_{i_2}(2)\cdots c_{i_k}(k)t^{i_0+i_1+\cdots+i_k}$ are all distinct.

15.6 Definition. A sequence $p_0, p_1, p_2, \ldots, p_k, \ldots$ of trigonometric polynomials is said to be dissociated if for each k the polynomials $p_0, p_1, p_2, \ldots, p_k$ are dissociated.

Classical Riesz Products and a Theorem of Peyriére

15.7. Consider now a sequence of trigonometric polynomials obtained as follows: Let $n_0 < n_1 < n_2 < \cdots$ be a sequence of positive integers such that for all $j \geq 1$,

$$n_j > 2(n_{j-1} + \cdots + n_0) \tag{1}$$

Let

$$p_j(t) = 1 + c_j t^{n_j} + \overline{c_j} t^{-n_j} = 1 + 2\Re(c_j t^j), \ t \in S^1, \ j = 0, 1, 2, \ldots.$$

We require that $|c_j| \leq 1/2$.

It is easy to see that for each k, the polynomials $p_1, p_2, \ldots, p_k$ are dissociated in view of condition (1) above, which we call the dissociation condition on the polynomials $p_j(t), j = 0, 1, 2, \ldots$. The finite product of such p_j's expands into a trigonometric polynomial:

$$\prod_{j=0}^{k} p_j(t) = \sum_{-1 \leq \varepsilon_0, \varepsilon_1, \ldots, \varepsilon_k \leq 1} b_{\varepsilon_0} b_{\varepsilon_1} \cdots b_{\varepsilon_k} t^{\varepsilon_0 n_0 + \varepsilon_1 n_1 \cdots \varepsilon_k n_k}$$

$$= \sum_{i=-N_k}^{N_k} d_i t^i, \quad N_k = n_0 + n_1 + n_2 + \cdots + n_k,$$

where d_i is zero if i is not of the form $\varepsilon_0 n_0 + \varepsilon_1 n_1 + \cdots + \varepsilon_k n_k$ ($\varepsilon_i = -1, 0,$ or 1), and if i is of the form $\varepsilon_0 n_0 + \varepsilon_1 n_1 + \cdots + \varepsilon_k n_k$ then d_i is $b_{\varepsilon_0} \cdot b_{\varepsilon_1} \cdots b_{\varepsilon_k}$ where $b_{\varepsilon_i} = c_i$ if $\varepsilon_i = 1$, $b_{\varepsilon_i} = 1$ if $\varepsilon_i = 0$, and $b_{\varepsilon_i} = \overline{c_i}$ if $\varepsilon_i = -1$. The formal expansion of the infinite product

$$\prod_{j=0}^{\infty} p_j(t) \tag{R}$$

results in the trigonometric series

$$\sum_{i=-\infty}^{\infty} d_i t^i, \tag{2}$$

with

$$\prod_{i=0}^{k} p_j(t) = \sum_{i=-N_k}^{N_k} d_i t^i = (N_k)\text{th partial sum of the series (2)}. \tag{3}$$

Let P_k denote the partial product (3) above. Then the measures $P_k dt, k = 1, 2, 3, \ldots$, are all non-negative with total measure one, hence have a subsequence which converges weakly to a probability measure μ. But in view of (1) above the measures $P_k dt$ themselves converge to μ and the Fourier series of μ is the series (2) which is also the formal expansion $\prod_{k=0}^{\infty} p_k$. We therefore say that the product (R) represents the measure μ.

15.8 Definition. The product (R) is called the classical Riesz product. It is identified with the associated measure μ or its Fourier series.

Note that the classical Riesz product (R) depends on two parameters: the sequence $n_k, k = 0, 1, 2, \ldots$, (with $n_k > 2(n_{k-1} + n_{k-2} + \cdots + n_0)$) and the sequence $c_k, k = 0, 1, 2, \ldots$. If the first parameter is fixed and if c stands for the second parameter we denote the classical Riesz product by μ_c to indicate its dependence of c. We have the following theorem due to J. Peyriére [24].

15.9 Theorem. *Let* $a = (a_k)_{k=0}^{\infty}$ *and* $b = (b_k)_{k=0}^{\infty}$ *be two sequences of complex numbers such that for each* k

$$|a_k| \leq 1/2, \quad |b_k| \leq 1/2, \quad \sum_{k \geq 0} |b_k - a_k|^2 = \infty.$$

Then the two measures μ_a *and* μ_b *are mutually singular.*

Proof. A calculation shows that the functions $t^{n_k} - \overline{a_k}, k = 0, 1, 2, \ldots$, form an orthogonal system in $L^2(S^1, \mu_a)$. Also,

$$\int_{S^1} |t^{n_k} - a_k|^2 \, d\mu_a = 1 - |a_k|^2 \leq 1.$$

Since $\sum_{k\geq 0} \mid b_k - a_k \mid^2$ is infinite there exists a sequence of complex numbers, $(\alpha_k)_{k\geq 0}$, which is square summable and such that:

(i) for each $k \geq 0$, $\alpha_k(\overline{b_k} - \overline{a_k})$ is non-negative,

(ii) $\sum_{k\geq 0} \alpha_k(\overline{b_k} - \overline{a_k}) = \infty$.

(This is an easy consequence of the Banach-Steinhaus theorem. For if such a sequence does not exist then for every square summable sequence $\alpha = (\alpha_k)_{k=0}^{\infty}$, the series $\sum_{k\geq 0}(b_k - a_k)\alpha_k$ converges to a finite complex number, say $l(\alpha)$. Thus l is a linear functional defined on all of l^2. In addition l can be seen to be the weak limit of a sequence of bounded linear functionals, hence by the Banach-Steinhaus theorem, l is also a bounded linear functional. Clearly, by the Riesz representation theorem, the sequence $(b_k - a_k)_{k=0}^{\infty}$ is square summable, contrary to the assumption.)

The series $\sum_{k\geq 0} \alpha_k(t^{n_k} - \overline{a_k})$ and $\sum_{k\geq 0} \alpha_k(t^{n_k} - \overline{b_k})$ are convergent in $L^2(S^1, \mu_a)$ and $L^2(S^1, \mu_b)$ respectively. There exists a subsequence $n_{k_l} = N_l$, $l = 0, 1, 2, \ldots$ of the sequence n_k, $k = 0, 1, 2, \ldots$, such that, as $l \to \infty$, the partial sums $\sum_{k=0}^{N_l} \alpha_k(t^{n_k} - \overline{a_k})$, $l = 1, 2, 3, \ldots$, converge a.e. μ_a and the partial sums $\sum_{k=0}^{N_l} \alpha_k(t^{n_k} - \overline{b_k})$ converge a.e. μ_b. If μ_a and μ_b are not mutually singular then there exists an $s \in S^1$ such that the two preceding partial sums converge when we substitute s for t. Taking the difference of these partial sums we see that the series of positive terms $\sum_{k\geq 0} \alpha_k(\overline{b_k} - \overline{a_k})$ converges, contrary to the hypothesis. This proves the theorem.

15.10 Corollary. *The classical Riesz product μ_a is absolutely continuous or singular to the Lebesgue measure according as $\sum_{n\geq 0} \mid a_k \mid^2$ is finite or infinite.*

Proof. If $b_k = 0$ for all k then μ_b is the Lebesgue measure so that if $\sum_{k\geq 0} \mid a_k \mid^2 = \infty$ then μ_a singular to Lebesgue measure by the above theorem. On the other hand if $\sum_{k\geq 0} \mid a_k \mid^2 < \infty$ then it can be verified that the sum of the squares of the Fourier coefficients of μ_a is finite too, indeed it is $\leq \prod_{k=0}^{\infty}(1 + \mid a_k \mid^2)$, so that μ_a is absolutely continuous with respect to the Lebesgue measure.

15.11. The dissociation condition $\forall k$, $n_{k+1} > 2(n_k + \cdots + n_0)$ is satisfied whenever $\forall k$, $\frac{n_{k+1}}{n_k} > 3$. Further if the c_k's are real and if $t = e^{ix}$ then $p_k(t) = 1 + 2c_k \cos n_k x = 1 + \alpha_k \cos n_k x$, where $\alpha_k = 2c_k$. Since $\mid c_k \mid \leq 1/2$, $\mid \alpha_k \mid \leq 1$. The product $\prod_{j=0}^{\infty} p_j(t)$ takes the form $\prod_{j=0}^{\infty}(1 + \alpha_j \cos n_j x)$, $-1 \leq \alpha_j \leq 1$, which is the classical Riesz product discussed in Zygmund [26].

Riesz Products and Dynamics

15.12. We will now show that every measure μ defined by a classical Riesz product appears as the maximal spectral type of the unitary operator associated with a non-singular dynamical system and a cocycle. (See B. Host, J.-F. Méla,

F. Parreau [14]). Indeed dynamics will allow us to show that the product (R) defines a measure under a weaker condition:

$$n_{k+1} > n_k + n_{k-1} + \cdots + n_0.$$

which is implied by the condition: $n_{k+1} > 2n_k$. This condition does not ensure that the polynomials $p_j, j = 0, 1, 2, \ldots$ are dissociated. The product (R) under this weaker condition on the n_j's will be called Riesz product, so that the words "classical Riesz product" will mean a Riesz product whose polynomials are dissociated.

15.13. Let $\Omega_0 = \{0, 1\}^{\aleph_0} =$ set of all sequences $(\omega_0, \omega_1, \omega_2, \cdots)$ of zeros and ones. We identify Ω_0 with the group of diadic integers; the group operation being addition coordinatewise (mod 2) with carry to the right. Let 1 denote the element $(1, 0, 0, \cdots) \in \Omega_0$. Let S be the automorphism defined on Ω_0 by $S\omega = \omega + 1$. We call S the diadic adding machine or the odometer. Let ν_j denote the measure on $\{0, 1\}$ given by $\nu_j(0) = p_{0,j}$, $\nu_j(1) = p_{1,j}$, $0 \le p_{0,j}, p_{1,j} \le 1$, $p_{0,j} + p_{1,j} = 1$. Let ν be the product measure $\prod_{j=0}^{\infty} \nu_j$. The measure ν is quasi-invariant and ergodic under S. Moreover ν is non-atomic if and only if $\sum_{j=0}^{\infty} \min(p_{0,j}, p_{1,j}) = \infty$. From now on we will assume that this condition holds.

15.14. Next let h be a non-negative integer valued function on Ω_0. Let $X \subseteq \Omega_0 \times \{0, 1, 2, 3, \ldots\}$ be the set of points (ω, n) with $0 \le n \le h(\omega)$, the part of $\Omega_0 \times \{0, 1, 2, \ldots\}$ below and including the graph of h. Define T on X by

$$T(\omega, n) = \begin{cases} (\omega, n+1) & \text{if } 0 \le n < h(\omega) \\ (S\omega, 0) & \text{if } n = h(\omega). \end{cases}$$

We know that X is the disjoint union of sets $X_n = \{(\omega, n) : h(\omega) = n\}$, $n = 0, 1, 2, 3, \ldots$. We define a measure μ on X by requiring that the restriction of μ to X_n, for each n, be given by

$$\mu(A) = \nu(\{\omega : (\omega, n) \in A\}), A \subseteq X_n.$$

The measure μ is σ-finite, quasi-invariant, and ergodic under T (because ν has the same properties under S). The automorphism T on (X, μ) is the automorphism built under h on the base space (Ω_0, ν) with the base automorphism S. For our purpose the sets of constancy of h have to be chosen suitably. This is described next.

15.15. We choose the sets of constancy of h as follows: For $\omega \in \Omega_0$, denote by $s(\omega)$ the smallest k such that $\omega_k = 0$, $(\omega = (\omega_0, \omega_1, \omega_2, \ldots))$. Let S_k denote the set of $\omega \in \Omega_0$ for which $s(\omega) = k$. The sets $S_k, k = 0, 1, 2, \ldots$, partition Ω_0 except that the point $(1, 1, 1, \ldots)$ does not belong to any of the sets S_k. We omit the S-orbit of this point from our consideration. Since the measure ν is free of atoms this will not matter to us. The function h is any non-negative

integer valued function whose sets of constancy are the sets $S_n, n = 0, 1, 2, \ldots$.
For each n, let a_n denote the value of h on S_n. Write $n_0 = a_0 + 1, n_1 = 2n_0 + a_1 + 1, \ldots, n_k = 2n_{k-1} + a_k + 1, \ldots$. Note that for $k \geq 1$, n_k is the first
return time of $\omega \in S_k$ into S_{k+1}.

15.16. We now define a measurable function ϕ of absolute value one on X
which will provide us with a $\mathbb{Z} \times X$ cocycle. Define

$$\phi(\omega, n) = \begin{cases} 1 & \text{if } n > 0, \\ c_k & \text{if } n = 0 \text{ and } \omega \in S_k, \end{cases}$$

where $c_0, c_1, c_2, \ldots$, are complex numbers of absolute value one.

15.17. Let T be as in **15.14.** (with h as in **15.15.**) and let ϕ be as above. Define
$U = U_T$ and $V = V_\phi$ on $L^2(X, \mathcal{B}, \mu)$ as follows:

$$(Uf)(x) = \left(\frac{d\mu_T}{d\mu}(x) \right)^{1/2} f(Tx),$$

$$(Vf)(x) = \phi(x) \left(\frac{d\mu_T}{d\mu}(x) \right)^{1/2} f(Tx) = \phi(x) \cdot (Uf)(x), \ f \in L^2(X, \mathcal{B}, \mu).$$

15.18 Theorem. *The operator V has simple spectrum. Its maximal spectral type
is given by the Riesz product*

$$\prod_{j=0}^{\infty} \left(1 + \sqrt{(p_{0,j})} \sqrt{(p_{1,j})} (\alpha_j t^{n_j} + \overline{\alpha_j} t^{-n_j}) \right)$$

*where for each j, α_j is a constant of absolute value one depending only on
$c_0, c_1, \ldots, c_j$. Every Riesz product (hence also every classical Riesz product)
appears as the maximal spectral type of a suitable V (up to a discrete measure).*

The proof needs a careful calculation. Although it is possible to make the
required calculations using the above description of T and ϕ, they are valid
more generally. (See J. R. Choksi and M. G. Nadkarni [7].) We will therefore
make these calculations in the more general setting of rank one automorphism
and then specialise to the above case. First we set forth below some basic facts
about generalised Riesz products.

Generalised Riesz Products

15.19 Definition. Let $P_1, P_2, \ldots$ be a sequence of trigonometric polynomials
such that

(i) for any finite sequence $i_1, i_2, i_3, \ldots, i_k$ of natural numbers

$$\int_{S^1} |\, (P_{i_1} P_{i_2} \cdots P_{i_k})(t) \,|^2 \, dt = 1,$$

(ii) for any infinite sequence $i_1 < i_2 < \cdots$, of natural numbers the weak limit of the measures $\mid (P_{i_1} P_{i_2} \cdots P_{i_k})(t) \mid^2 dt$ as $k \to \infty$ exists. Then the measure μ given by the weak limit of $\mid (P_1 P_2 \cdots P_k)(t) \mid^2 dt$ as $k \to \infty$ is called the generalised Riesz product of the polynomials $\mid P_1 \mid^2, \mid P_2 \mid^2, \ldots$, and denoted by $\prod_{j=1}^{\infty} \mid P_j \mid^2$.

15.20 Remark 1. In general the weak limit of $\mid (P_{i_1} P_{i_2} \cdots P_{i_k})(t) \mid^2 dt$, (which is a generalised Riesz product in its own right), depends on the sequence $i_1 < i_2 < i_3 < \cdots$.

Remark 2. The generalised Riesz products that we will encounter in this chapter will have the additional property that $\mu_k \to$ Haar measure on S^1 weakly, where

$$\mu_k = \prod_{j=k+1}^{\infty} \mid P_j \mid^2 .$$

Remark 3. Generalised Riesz Products are briefly mentioned in B. Host, J.-F. Méla, F. Parreau [14] where references to earlier work can be found. The expanded definition given above is suggested by the considerations in I. Klemes and K. Reinhold [19].

15.21 Exercise 1. Let $P_j, j = 0, 1, 2, \ldots$, be a sequence of trigonometric polynomials such that for all j, $\int_{S^1} \mid P_j(t) \mid^2 dt = 1$ and the polynomials $\mid P_j \mid^2, j = 0, 1, 2, \ldots$, are dissociated. Let $i_1 < i_2 < \cdots$, be a sequence of natural numbers. Show that $\mid (P_{i_1} P_{i_2} \cdots P_{i_k})(t) \mid^2 dt$ are all probability measures whose weak limit as $k \to \infty$ exists. In other words the generalised Riesz product $\prod_{j=1}^{\infty} \mid P_j \mid^2$ exists.

15.22 Exercise 2. Assume that the sequences of trigonometric polynomials $P_j(z) = \sum_{k=0}^{N_j} a_{k,j} z^{n_{k,j}}$, $Q_j(z) = \sum_{k=0}^{N_j} b_{k,j} z^{n_{k,j}}$, $j = 1, 2, \ldots$ are such that

(i) for each j, the integers $n_{k,j} - n_{l,j}$, $k \neq l$ are all distinct,
(ii) for all j, $\int_{S^1} \mid P_j(t) \mid^2 dt = 1$ and $\mid P_j \mid^2, j = 1, 2, \ldots$, are dissociated,
(ii) for all j, $\int_{S^1} \mid Q_j(t) \mid^2 dt = 1$ and $\mid Q_j \mid^2, j = 1, 2, \ldots$, are dissociated,
(iii) $\sum_{j=1}^{\infty} \sum_{k=1}^{N_j} \mid a_{0,j} a_{k,j} - b_{0,j} b_{k,j} \mid^2 = \infty$.

Show that the generalised Riesz products $\prod_{j=1}^{\infty} \mid P_j \mid^2$, $\prod_{j=1}^{\infty} \mid Q_j \mid^2$ are mutually singular.

The following proposition (shown to me by F. Parreau) gives a criterion for two generalised Riesz products, which are not necessarily dissociated, to be mutually singular.

15.23 Proposition. *Let* $\mu = \prod_{j=1}^{\infty} \mid P_j \mid^2, \nu = \prod_{j=1}^{\infty} \mid Q_j \mid^2$ *be two generalised Riesz products. Let*

$$\mu_n = \prod_{j=n+1}^{\infty} \mid P_j \mid^2, \quad \nu_n = \prod_{j=n+1}^{\infty} \mid Q_j \mid^2 .$$

Assume that

$$\prod_{j=1}^{n} |P_j|^2 \, d\nu_n \to \mu \quad \text{weakly as } n \to \infty,$$

$$\prod_{j=1}^{n} |Q_j|^2 \, d\mu_n \to \nu \quad \text{weakly as } n \to \infty.$$

Then the following are equivalent:

(a)

$$\inf_{n \in \mathbb{N}} \int_{S^1} \left| \frac{Q_1 \cdot Q_2 \cdots Q_n}{P_1 \cdot P_2 \cdots P_n} \right| d\mu = 0,$$

(b) *μ and ν are mutually singular.*

Proof. $(a) \Rightarrow (b)$. Let

$$f = \left(\frac{d\nu}{d\mu} \right)^{1/2} = \left(\prod_{j=1}^{n} \left| \frac{Q_j}{P_j} \right| \right) f_n,$$

where

$$f_n = \left(\frac{d\nu_n}{d\mu_n} \right)^{1/2}; \quad f^2 = \frac{d\nu}{d\mu} = \left(\prod_{j=1}^{n} \left| \frac{Q_j}{P_j} \right|^2 \right) \frac{d\nu_n}{d\mu_n}.$$

Now

$$0 \le \int_{S^1} f d\mu = \int_{S^1} \prod_{j=1}^{n} \left| \frac{Q_j}{P_j} \right| f_n d\mu$$

$$\le \left(\int_{S^1} \prod_{j=1}^{n} \left| \frac{Q_j}{P_j} \right| d\mu \right)^{1/2} \cdot \left(\int_{S^1} \prod_{j=1}^{n} \left| \frac{Q_j}{P_j} \right| f_n^2 d\mu \right)^{1/2}$$

(this is obtained by applying Schwarz inequality to the functions 1 and f_n with respect to the measure $\prod_{j=1}^{n} | \frac{Q_j}{P_j} | d\mu$),

$$\le \left(\int_{S^1} \prod_{j=1}^{n} \left| \frac{Q_j}{P_j} \right| d\mu \right)^{1/2} \cdot \left(\int_{S^1} \prod_{j=1}^{n} \left| \frac{Q_j}{P_j} \right|^2 f_n^2 d\mu \right)^{1/4} \cdot \left(\int_{S^1} f_n^2 d\mu \right)^{1/4}$$

(this is obtained by applying Schwarz inequality to the functions 1 and $\prod_{j=1}^{n} | \frac{Q_j}{P_j} |$ with respect to the measure $f_n^2 d\mu$)

$$\le \left(\int_{S^1} \prod_{j=1}^{n} \left| \frac{Q_j}{P_j} \right| d\mu \right)^{1/2} \cdot \left(\int_{S^1} f^2 d\mu \right)^{1/4} \cdot \left(\int_{S^1} d\nu \right)^{1/4}.$$

Since the last two terms of this product remain bounded away from infinity as $n \to \infty$ we see from (a) that $f = 0$ *a.e.* ν, whence μ and ν are mutually singular.

$(b) \Rightarrow (a)$. Assume now that (b) holds and that

$$\inf_{n \in \mathbb{N}} \int_{S^1} \left| \frac{Q_1 \cdot Q_2 \cdots Q_n}{P_1 \cdot P_2 \cdots P_n} \right| d\mu > 0.$$

We arrive at a contradiction as follows:

Since $\| \prod_{j=1}^{n} \frac{Q_j}{P_j} \|_{L^2(S^1,\mu)} \to \int_{S^1} d\nu = 1$, the collection $\prod_{j=1}^{n} | \frac{Q_j}{P_j} |$, $n = 1, 2, \ldots$, has an L^2 weak limit, say ξ. Since

$$\inf_{n \in \mathbb{N}} \int_{S^1} \left| \frac{Q_1 \cdot Q_2 \cdots Q_n}{P_1 \cdot P_2 \cdots P_n} \right| d\mu > 0,$$

the limit is non-zero. Since μ is singular to ν, we can choose a continuous $\phi \geq 0$ such that

(i) $\int_{S^1} \phi \xi d\mu = C > 0$
(ii) $\int_{S^1} \phi d\mu = 1$
(iii) $\int_{S^1} \phi d\nu < \varepsilon < C^2$

Now

$$C = \lim_{n \to \infty} \int_{S^1} \prod_{j=1}^{n} \left| \frac{Q_j}{P_j} \right| \phi d\mu$$

$$\leq \left(\lim_{n \to \infty} \int_{S^1} \prod_{j=1}^{n} \left| \frac{Q_j}{P_j} \right|^2 \phi d\mu \right)^{1/2} \left(\int_{S^1} \phi d\mu \right)^{1/2}$$

$$\leq \lim_{n \to \infty} \left(\int_{S^1} \prod_{j=1}^{n} | Q_j |^2 \phi d\mu_n \right)^{1/2}$$

$$= \left(\int_{S^1} \phi d\nu \right)^{1/2} < \varepsilon^{1/2} < C,$$

a contradiction. This proves the proposition.

Two observations of J. Bourgain [6] follow as corollaries:

15.24 Corollary 1. *A generalised Riesz product $\mu = \prod_{j=1}^{\infty} | P_j |^2$ with $\mu_n \to dt$ as $n \to \infty$ is singular to Haar measure if and only if*

$$\inf_{n \in \mathbb{N}} \int_{S^1} \prod_{j=1}^{n} | P_j(t) | \, dt = 0.$$

Proof. If we take $Q_j = 1$ for every j, then ν is the Haar measure on S^1. The hypothesis of the proposition is satisfied (with the role of P's and Q's reversed) hence the conclusion holds.

Corollary 2. *A generalised Riesz product $\mu = \prod_{j=1}^{\infty} \mid P_j \mid^2$ with $\mu_n \to dt$ as $n \to \infty$ is singular to Haar measure if for some subsequence $k_1 < k_2 < \cdots$, of natural numbers*

$$\inf_{n \in \mathbb{N}} \int_{S^1} \prod_{j=1}^{n} \mid P_{k_j}(t) \mid dt = 0.$$

Proof. Let $A_n = \{1, 2, \ldots n\} \cap \{k_1, k_2, \ldots\}$, $B_n = \{1, 2, \ldots, n\} - A_n$. Then

$$\int_{S^1} \prod_{j=1}^{n} \mid P_j(t) \mid dt$$

$$\leq \left(\int_{S^1} \prod_{j \in B_n} \mid P_j(t) \mid^2 \cdot \prod_{j \in A_n} \mid P_j(t) \mid dt \right)^{1/2} \cdot \left(\int_{S^1} 1 \cdot \prod_{j \in A_n} \mid P_j(t) \mid dt \right)^{1/2}$$

$$\leq \left(\int_{S^1} \prod_{j=1}^{n} \mid P_j(t) \mid^2 dt \right)^{1/4} \cdot \left(\int_{S^1} 1 \cdot \prod_{j \in A_n} \mid P_j(t) \mid dt \right)^{1/2}$$

$$\leq \left(\int_{S^1} 1 \cdot \prod_{j \in A_n} \mid P_j(t) \mid dt \right)^{1/2} .$$

The corollary follows.

Maximal Spectral Types of Rank One Automorphisms

15.25. The maximal spectral type of a general rank one automorphism is given by a generalised Riesz product with possibly some discrete measure. For a weakly mixing rank one automorphism or for an infinite measure preserving rank one automorphism the description of the maximal spectral type is exact. We will assume that the reader is familiar with the method of cutting and stacking to construct rank one automorphisms.

15.26. Divide the unit interval Ω_0 into m_1 equal parts, add spacers, and form a stack of a certain height (say h_1) in the usual fashion. This is the first stage of our construction. At the k-th stage, divide the stack obtained at the $(k-1)$-th stage into m_k equal parts, add spacers, and obtain a new stack (say of height h_k) in the usual fashion. If during the k-th stage of construction the number of spacers put above the j-th column of the $(k-1)$-th stack is $a_j^{(k)}$, $0 \leq a_j^{(k)} < \infty$, $1 \leq j \leq m_k$, then we have

$$h_k = m_k h_{k-1} + \sum_{j=1}^{m_k} a_j^{(k)}.$$

Proceeding thus we get a rank one automorphism T on a certain measure space $(X, \mathcal{B}, m)$ which may be finite or σ-finite depending on the number of spacers added. Since we are concerned with general rank one automorphisms, no control is necessary over the manner in which the spacers are added except that we add only finitely many spacers at each stage.

15.27. For each $k = 1, 2, 3, \ldots$, let Ω_k denote the base of the stack at the end of the k-th stage of construction. Note that

$$m(\Omega_k) = \frac{1}{m_1} \frac{1}{m_2} \cdots \frac{1}{m_k}.$$

For $A, B \subseteq X$, write $R_i(x, A, B)$ to denote the i-th entry time into B of the point $x \in A$, with the convention that $R_1(x, A, B) > 0$ even if $x \in A \cap B$. Note that $R_i(x, \Omega_k, \Omega_{k-1})$ is independent of $x \in \Omega_k$ for $i = 1, 2, \ldots, m_k - 1$. We therefore write $R_i(x, \Omega_k, \Omega_{k-1}) = R_{i,k}$, $i = 1, 2, \ldots, m_k - 1$. Note that

$$R_{i,k} = i h_{k-1} + a_1^{(k)} + \cdots + a_i^{(k)}, \ 1 \leq i \leq m_k - 1.$$

We have

$$\Omega_{k-1} = \Omega_k \cup T^{R_{1,k}} \Omega_k \cup \cdots \cup T^{R_{m_k-1,k}} \Omega_k$$

$$1_{\Omega_{k-1}} = P_k(U) 1_{\Omega_k} \tag{1}$$

where

$$P_k(U) = I + U^{-R_{1,k}} + \cdots + U^{-R_{m_k-1,k}}$$

and U is the unitary operator on $L^2(X, \mathcal{B}, m)$ defined by $Uf = f \circ T$, $f \in L^2(X, \mathcal{B}, m)$. Note that in terms of the stack heights and spacer lengths

$$P_k(U) = I + \sum_{j=1}^{m_k-1} U^{-(jh_{k-1} + a_1^{(k)} + \cdots + a_j^{(k)})}.$$

Iterating (1) we get

$$1_{\Omega_0} = \left(\prod_{j=1}^{k} P_j(U) \right) 1_{\Omega_k}.$$

Let us normalise 1_{Ω_k} and write $f_k = (m(\Omega_k))^{-1/2} 1_{\Omega_k}$, $k = 0, 1, 2, \ldots$. Note that $m(\Omega_0) = 1$ so that $f_0 = 1_{\Omega_0}$. We have

$$f_0 = \left[(m(\Omega_k))^{1/2} \prod_{j=1}^{k} P_j(U) \right] f_k.$$

If σ_k denotes the measure on S^1 defined by

$$(U^n f_k, f_k) = \int_{S^1} z^n d\sigma_k, \ \ n \in \mathbb{Z}, \ k = 0, 1, 2, 3, \ldots$$

then we see that

$$d\sigma_0 = \left(\prod_{j=1}^k \mid P_j(z) \mid^2 \right) \cdot m(\Omega_k) d\sigma_k, \ \ k = 1, 2, 3, \ldots,$$

or

$$d\sigma_0 = \left(\prod_{j=1}^k \frac{1}{m_j} \mid P_j(z) \mid^2 \right) d\sigma_k, \ \ k = 1, 2, 3, \ldots. \tag{2}$$

Since the co-efficient of $d\sigma_k$ can vanish only at finitely many points (being a trigonometric polynomial) we see that the non-atomic parts of σ_0 and σ_k are mutually absolutely continuous. Moreover $\vee_{k=1}^\infty \sigma_k$ is the maximal spectral type of U, whence σ_0 and the maximal spectral type σ of U have their non-atomic part mutually absolutely continuous, i.e., in the same measure class.

15.28. Let us replace $d\sigma_k$ by dz in the right hand side of (2) above and let ρ_k denote the resulting measure:

$$d\rho_k = \left(\prod_{j=1}^k \frac{1}{m_j} \mid P_j(z) \mid^2 \right) dz, \ \ k = 1, 2, 3, \ldots.$$

It is easy to see that $\sigma_k \to dz$ as $k \to \infty$, hence it seems natural to replace $d\sigma_k$ by dz in the right-hand side of (2) and to surmise that σ_0 is the weak limit of the $d\rho_k$'s. This is indeed the case as we see below. We have:

15.29 Theorem. The measure σ_0 is the weak limit of the $\rho_k, k = 1, 2, 3, \ldots$. In other words, for each $n \in \mathbb{Z}$ we have $\hat{\rho}_k(n) \to \hat{\sigma}_0(n)$ as $k \to \infty$.

Proof. let N_k denote the set of integers consisting of zero together with the entry times $R_1(x, \Omega_k, \Omega_0), R_2(x, \Omega_k, \Omega_0), \ldots$, which are less than the height h_k of the k-th stack. We note that under this constraint $R_i(x, \Omega_k, \Omega_0)$ is independent of $x \in \Omega_k$. We further have

$$\Omega_0 = \cup_{s \in N_k} T^s \Omega_k, \ \ \text{a disjoint union,}$$

$$1_{\Omega_0} = \sum_{s \in N_k} U^{-s} 1_{\Omega_k},$$

$$f_0 = \left(\sum_{s \in N_k} U^{-s} f_k \right) (m(\Omega_k))^{1/2} = Q_k(U) f_k$$

where

$$Q_k(U) = \left(\sum_{s \in N_k} U^{-s} \right) (m(\Omega_k))^{1/2}.$$

Clearly $Q_k(z) = \left(\prod_{j=1}^{k} P_j(z) \right) (m(\Omega_k))^{1/2}$ and $| Q_k(z) |^2 = \frac{d\rho_k}{dz}$.

Now fix $n \in \mathbb{Z}$ and let k be so large that the first return time of any $x \in \Omega_k$ back to Ω_k (under T or T^{-1}) is bigger than $| n |$; equivalently, let k be so large that $h_k > | n |$. If $r, s \in N_k$ then $s + n - r$ can never exceed or equal the second return time of an $x \in \Omega_k$ back to Ω_k (under T or T^{-1}). Moreover there are at most n^2 pairs (r, s) with $r, s \in N_k$ such that $s + n - r$ equals the first return time of an $x \in \Omega_k$ back to Ω_k. For suppose $n > 0$ and $T^{s+n-r}\Omega_k \cap \Omega_k \neq \emptyset$ and $s + n - r \neq 0$. Then $r = n + s - u$ where u is the first return time of some $x \in \Omega_k$ back to Ω_k. Since each n, r, s is less than h_k, $h_k \leq u$ and $r \geq 0$, we see that $0 \leq r < n$ and $s + n - r = u \geq h_k$, so $s \geq h_k - (n - r)$. Thus there can be at most n^2 pairs (r, s), $r, s \in N_k$, with $T^{s+n-r}\Omega_k \cap \Omega_k \neq \emptyset$ and $s + n - r \neq 0$. (But note that for each fixed u, there are at most n pairs (r, s) with this property.) A similar argument holds for $n < 0$. So, if $T^{s+n-r}\Omega_k \cap \Omega_k \neq \emptyset$, then we must have $s + n - r = 0$ except for at most n^2 pairs $(r, s), r, s \in N_k$. Now

$$(U^n f_0, f_0) = \sum_{r,s \in N_k} (U^{n-r} f_k, U^{-s} f_k) m(\Omega_k)$$

$$= \sum_{r,s \in N_k} (U^{s+n-r} 1_{\Omega_k}, 1_{\Omega_k})$$

$$= \sum_{r,s \in A} (U^{s+n-r} 1_{\Omega_k}, 1_{\Omega_k}) + \sum_{r,s \in B} (U^{s+n-r} 1_{\Omega_k}, 1_{\Omega_k})$$

where A is the set of pairs $(r, s), r, s \in N_k$ with $s + n - r = 0$ and B is the set of pairs $(r, s), r, s \in N_k$ with $s + n - r$ equal to the first return time of an $x \in \Omega_k$ back to Ω_k (under T or T^{-1}). Now

$$\sum_{r,s \in B} (U^{s+n-r} 1_{\Omega_k}, 1_{\Omega_k}) \leq n^2 m(\Omega_k) \to 0 \quad \text{as} \quad k \to \infty,$$

whence

$$(U^n f_0, f_0) = \lim_{k \to \infty} L_k \cdot m(\Omega_k),$$

where L_k is the number of pairs $(r, s), r, s \in N_k$ with $s + n - r = 0$. (By breaking up Ω_k into disjoint sets corresponding to each first return time u we could even replace the term $n^2 m(\Omega_k)$ by $| n | m(\Omega_k)$.) On the other hand, it is easy to see that

$$\int_{S^1} z^n d\rho_k = \int_{S^1} z^n m(\Omega_k) \left| \sum_{s \in N_k} z^{-s} \right|^2 dz = m(\Omega_k) \cdot L_k,$$

so that $\hat{\rho}_k(n) \to \hat{\sigma}_0(n)$ as $k \to \infty$, for each $n \in \mathbb{Z}$, and so σ_0 is the weak limit of ρ_k. This proves the theorem. (For an alternative proof, not involving dynamics, see I. Klemes and K. Reinhold [19].)

Examples and Remarks

15.30 Example (a). Consider Chacon's automorphism (see N. Friedman [12]), where at each stage we divide the stack into three equal parts and place a single spacer on the top of the middle column. We have

$$h_1 = 3 + 1,$$

$$h_2 = 3(3+1) + 1,$$

$$\vdots$$

$$h_n = 3^n + 3^{n-1} + \cdots + 1 = \frac{3^{n+1} - 1}{2}.$$

The ρ_k's take the form

$$d\rho_k = \left(\frac{1}{3^k} \prod_{j=1}^{k} \left| \left(1 + z^{-h_j} + z^{-(2h_j+1)} \right) \right|^2 \right) dz.$$

In other words the maximal spectral type of Chacon's automorphism is given by the generalised Riesz product

$$\prod_{j=1}^{\infty} \frac{1}{3} \left| \left(1 + z^{-\frac{3^{j+1}-1}{2}} + z^{-3^{j+1}} \right) \right|^2.$$

This automorphism is weakly mixing but not mixing and has no square roots, so can not be embedded in a flow. Moreover it is prime in the sense that it does not admit any non-trivial invariant sub-σalgebras. (See A. del Junco, M. Rahe, L. Swanson [10]). Further, it has singular spectrum; this was first shown by J. Baxter [2] but now follows on applying the criterion for singularity of a generalised Riesz product.

15.31. The question whether there exist weak mixing automorphisms without square roots was solved in the affirmative by R. Chacon [7]. D. Ornstein [22] answered the same question for mixing automorphisms (see chapter **16**). It is a rank one construction in which the spacers at each stage are added in a certain random manner and $m_k \to \infty$ speedily so that for each ω in a certain probability space Ω we obtain a measure preserving automorphism T_ω such that for a.e. $\omega \in \Omega$, T_ω is mixing, and has no square roots (indeed it commutes only with its powers). It was also shown by Ornstein that these automorphisms are prime in the sense that they do not admit non-trivial invariant sub-σ-algebras. It was subsequently proved by del Junco (see [10]) that the simpler Chacon's automorphism also has these properties. A careful study of Ornstein's paper has lead to the introduction of new methods (e.g., the machinery of self-joinings) in ergodic theory. The nature of the spectrum of U_{T_ω} remained unknown until J. Bourgain proved that for a.e. $\omega \in \Omega$, U_{T_ω} has singular spectrum. Recently El

Houcein [15] has shown that for a.e. pair (ω, ω'), U_{T_ω} and $U_{T_{\omega'}}$ have mutually singular spectra.

15.32 Example (b). The staircase automorphism. Here at the k-th stage, we divide the $(k-1)$-th stack into k equal columns and put j spacers over the j-th column, $1 \le j \le k$, (hence the name 'staircase'), and then stack. Note that at the first stage we do not divide Ω_0 at all, but only add a spacer equal to the length of Ω_0. We have

$$h_1 = 2, h_2 = 2 \times 2 + 1 + 2 = 7, \ldots, h_k = kh_{k-1} + \frac{k(k+1)}{2}$$

and the maximal spectral type of the staircase automorphism is given by the generalised Riesz product:

$$\prod_{k=1}^{\infty} |P_j(z)|^2,$$

where

$$P_j(z) = \frac{1}{\sqrt{j}}(1 + z^{-(h_j-1)+1} + z^{-(2h_j-1+1+2)} + \cdots + z^{-[(j-1)h_j-1+\frac{(j-1)j}{2}]}$$

The staircase automorphism is known to be mixing (T. Adams [1]) and to have singular spectrum (I. Klemes [18]).

15.33. The infinite product

$$\prod_{l=1}^{\infty}\left(\frac{1}{m_{j_l}}|P_{j_l}|^2\right)$$

taken over a subsequence $j_1 < j_2 < j_3 \cdots$, also represents the maximal spectral type (up to a discrete measure) of some rank one automorphism. In case $j_i \ne i$ for infinitely many i, then the automorphism acts on an infinite mesure space.

15.34. In case $m(X)$ is finite, σ_0 has a non-trivial mass at $z = 1$ so that $\sum_{n\in\mathbb{Z}} |\hat{\sigma}_0(n)|^2 = \infty$. It is interesting to note that this fact, viz., $\sum_{n\in\mathbb{Z}} |\hat{\sigma}_0(n)|^2 = \infty$, always holds, whether $m(X)$ is finite or not. Indeed, since the coefficients of powers of z in the formal expansion (i.e. without grouping terms) of the infinite product $\prod_{k=1}^{\infty} \frac{1}{m_k} |P_k(z)|^2$ are all positive and since $\hat{\sigma}_0(n) = $ sum of the coefficients z^n in this formal expansion, we see that $\sum_{n\in\mathbb{Z}} |\hat{\sigma}_0(n)|^2 \ge$ sum of the squares of the coefficients of the powers of z in the formal expansion of the infinite product. This second sum of squares in turn is bigger than $\sum_{k=1}^{\infty} \frac{m_k(m_k-1)}{m_k^2}$, a sum which is ∞. If m_k's are bounded over a subsequence then over a further subsequence the polynomials $\frac{1}{m_k} |P_k|^2$ are dissociated and the corresponding generalised Riesz subproduct represents a singular measure in view of its Fourier coefficients not being in L^2. It follows from Bourgain's observation that in such a case the original generalised Riesz product is itself singular.

The Non-Singular Case, Proof of Theorem 15.18, and Further Remarks

15.35. One can consider non-singular T obtained by cutting and stacking. This means that at the k-th stage we divide the stack obtained at the $(k-1)$-th stage in the ratios

$$p_{0,k}, p_{1,k}, \ldots, p_{m_k-1,k}, \quad p_{ik} > 0, \quad \sum_{i=0}^{m_k-1} p_{i,k} = 1.$$

The spacers are added in the usual manner, by which we mean that the sizes of the spacers added on the top of the j-th column are all the same and equal to the top piece of the j-th column. The extension of T to the spacers is done linearly as usual. Note that at the k-th stage the resultant measure is defined only for the algebra generated by the levels of the k-th stack. The resulting T after all stages of construction are complete is a non-singular ergodic automorphism for which $\frac{dm_T}{dm}$ is constant on all but the top layer of every stack. On $L^2(X, \mathcal{B}, m)$ we now define

$$(Vf)(x) = A(x) \cdot \left(\frac{dm_T}{dm}\right)^{1/2}(x)f(Tx), f \in L^2(X, \mathcal{B}, m),$$

where A is a function of absolute value one which is constant on all but the top layer of every stack and $m_T = m \circ T$. It can be shown by the above method that the maximal spectral type of V (up to a discrete measure) is given by the weak limit of the measures ρ_k defined as follows:

$$d\rho_k = \left(\prod_{j=1}^{k} p_{0,j} \mid P_j(z) \mid^2\right) dz,$$

where

$$P_j(z) = 1 + a_{1,j}\left(\frac{p_{1,j}}{p_{0,j}}\right)^{1/2} z^{-R_{1,j}} + \cdots + a_{m_j-1,j}\left(\frac{p_{m_j-1,j}}{p_{0,j}}\right)^{1/2} z^{-R_{m_j-1,j}}$$

and where $a_{1,j}, a_{2,j}, \ldots, a_{m_j-1,j}$ are constants of absolute value one determined by A.

For a precise and full details of this calculation see chapter 17, sections **17.23-17.27**.

15.36 Proof of Theorem 15.18. If $m_n = 2$ for all n the associated T is of the kind described in **15.14** (with h as in **15.15**) and the associated V as in theorem **15.18** with a slightly specialised cocycle A. The maximal spectral type of V is, up to a discrete measure, the Riesz product μ as stated in the theorem. Further given any Riesz product it is possible to choose the probabilities $p_{0,j}, p_{1,j}, \ j =$

$0, 1, 2, 3, \ldots$, and the constants $c_0, c_1, \ldots$, in such a way that the Riesz product associated with the resulting V is precisely μ. This proves the theorem.

15.37. Suppose $A = 1$ so that all $a_{i,j}$ are equal to one. Then it can be seen after a calculation that $\sum_{n \in \mathbb{Z}} | \hat{\sigma}_0(n) |^2 \geq \sum_{j=1}^{\infty} p_{0,j}(1 - p_{0,j})$, where

$$\hat{\sigma}_0(n) = (V^n 1_{\Omega_0}, 1_{\Omega_0}), \; n \in \mathbb{Z}.$$

If $\sum_{n \in \mathbb{Z}} | \hat{\sigma}_0(n) |^2$ is finite then $\sum_{j=1}^{\infty} p_{0,j}(1 - p_{0,j})$ is finite so that by ergodicity of T the measure on Ω_0 is discrete, whence the measure m on $\mathcal{B}$ (the σ-algebra generated by the intervals in the various stacks) is discrete.

15.38. Assume now that T is measure preserving. It is not known if T can be chosen so that U_T has spectrum absolutely continuous with respect to the Haar measure in S^1. More generally, it is not known if there exists a function A taking values $+1$ or -1 such that the associated V has absolutely continuous (with respect to the Haar measure on S^1) maximal spectral type. Mélanie Guenais [13] has shown that this is intimately related to the unresolved problem of the existence of 'flat polynomials' with co-efficient $+1$ and -1. A sequence of polynomials $P_n(x) = \sum_{j=0}^{p_n - 1} c_j(n) z^j, n = 1, 2, 3, \ldots$, with coefficients $c_j(n)$'s of unit modulus is called flat (with respect to the L^p norm, $p \neq 2$) if $\| P_n \|_p / \sqrt{p_n} \to 1$ as $n \to \infty$. Note that $\sqrt{p_n}$ is the L^2 norm of P_n, so that the sequence $P_n, n = 1, 2, 3, \ldots$, is flat if the L^2 and L^p metrics on these polynomials are equivalent. The construction of flat polynomials with respect to the L^1 norm has been set up by D. J. Newman in [21] and for the L^p-norm, $2 \leq p < \infty$ by E. Beller in [3]. J. -P. Kahane [17] eventually showed the existence of flat polynomial for $p = \infty$. All these constructions yield polynomials with complex coefficients and the problem of the existence of such polynomials with coefficients $+1$ and -1 remains open. It is shown in [13] that if such polynomials exist then there exists an automorphism with a simple Lebesgue component in its spectrum. It is shown further that on the compact group $X = \prod_{n \geq 0} \mathbb{Z}/p_n \mathbb{Z}$, where $p_n, n = 1, 2, 3, \ldots$, is a sequence of primes, there exist flat polynomials (with respect to the L^1-norm) with coefficients $+1$ and -1, and it is possible to construct an action of the group $\oplus_{n \geq 0} \mathbb{Z}/p_n \mathbb{Z}, \sum_{n=1}^{\infty} 1/\sqrt{p_n} < \infty$ which has a Lebesgue component of unit multiplicity.

15.39. There are aspects of the theory of Riesz products such as those considered in G. Brown and A. H. Dooley [4,5] which are not discussed in this chapter.

Rank One Automorphisms: Their Group of Eigenvalues

15.40. We will now compute the group $e(T)$ of L^∞ eigenvalues of a general rank one automorphism T. These will be the L^2 eigenvalues when the underlying space is of finite measure. Our expression for the eigenvalue group is intimately related to the corresponding expression for the maximal spectral type of T calculated in sections **15.28, 15.29**. Indeed for the T considered in these sections the group $e(T)$ is precisely the set of z for which the infinite product $\prod_{j=1}^{\infty} \frac{1}{m_j^2} \mid P_j(z) \mid^2$ converges to a finite non-zero value. This raises certain natural questions about the group of quasi-invariance of the maximal spectral type of T. We prove our results for measure preserving automorphisms, but they can be extended to non-singular automorphisms obtained by cutting and stacking.

15.41.   Descriptions of eigenvalue groups of certain non-singular flows were given by M. Osikawa [23] and by Y. Ito, T. Kamae and I. Shiokawa [16]. These authors were motivated by certain questions in non-singular weak equivalence theory. From the point of view of spectral theory, however, it is advantageous to recast their work using the "cutting and stacking" description of rank one automorphisms and some results on Fourier transforms of products of circle valued independent random variables (see J. R. Choksi and M. G. Nadkarni [9]).

Preliminary Calculations

15.42.   As before divide the unit interval Ω_0 into m_1 equal parts, add spacers and form a stack of height h_1 in the usual fashion. At the kth stage we divide the stack obtained at $(k-1)^{th}$ stage into m_k equal columns, add spacers and obtain a new stack of height h_k. If during the kth stage of our construction the number of spacers put above the jth column of the $(k-1)^{th}$ stack is $a_j^{(k)}$, $0 \le a_j^{(k)} < \infty$, $1 \le j \le m_k$, then we have

$$h_k = m_k h_{k-1} + \sum_{j=1}^{m_k} a_j^{(k)}.$$

Proceeding thus we get a rank one automorphism T on a certain measure space $(X, \mathcal{B}, m)$ which may be finite or σ-finite, depending on the number of spacers added. For each $k = 1, 2, 3, \ldots$, let Ω_k and Ω^k denote respectively the base and the top of the kth stack; of course $\Omega_k \subseteq \Omega_0$. There is no loss of generality in assuming in addition that $\Omega^k \subseteq \Omega_0$, i.e., no spacers are added on the last column at any stage in the construction. For given a rank one automorphism T constructed by cutting and stacking as above, we can construct as follows an isomorphic automorphism S with no spacers added on the last column at any stage: initially, cut Ω_0 into m_1 equal pieces, add $b_j^{(1)} = a_j^{(1)}$ spacers on the jth column, $1 \le j < m_1$, and stack. No spacers are added on the last column, i.e. $b_{m_1}^{(1)} = 0$. Cut Ω_1 into m_2 equal parts and add

$$b_j^{(2)} = a_j^{(2)} + a_{m_1}^{(1)}$$

spacers on the jth column $1 \leq j < m_2$ and stack; again $b_{m_2}^{(2)} = 0$. At the kth stage of the construction cut Ω_{k-1} into m_k equal pieces and add

$$b_j^{(k)} = a_j^{(k)} + \sum_{l=1}^{k-1} a_{m_l}^{(l)}$$

spacers on the jth column, $1 \leq j < m_k$, and stack; again $b_{m_k}^{(k)} = 0$. It is easily verified that the two automorphisms T and S with spacers $a_j^{(k)}$ and $b_j^{(k)}$ respectively are isomorphic, but no spacers are added on the last column at any stage in the construction of S. From now on we assume that $\Omega^k \subset \Omega_0$ for all k.

15.43. We denote the m_k equal columns obtained by dividing the $(k-1)^{th}$ stack by $C_1^k, \ldots, C_{m_k}^k$. For $1 \leq i \leq m_k$, write

$$Q_i^k = \text{union of parts of } \Omega_0 \text{ in the column } C_i^k.$$

Then $\{Q_1^k, \ldots, Q_{m_k}^k\}$ gives a partition $\mathcal{P}_k$ of Ω_0, and the partitions

$$\mathcal{P}_0, \mathcal{P}_1, \mathcal{P}_2, \ldots,$$

form an independent sequence of partitions of Ω_0; $\mathcal{P}_0$ being the trivial partition. They correspond to the partitions of the product space

$$\Omega = \prod_{k=1}^{\infty} \{0, 1, 2, \ldots, m_k - 1\}$$

given by the co-ordinate functions. Let τ denote the automorphism on Ω_0 induced by T. We know that τ is isomorphic to the odometer action on Ω.

The Functions γ_k

15.44. We now define a sequence γ_k, $k = 0, 1, 2, 3, \ldots$, of independent integer valued random variables on Ω_0. First define

$$\lambda_0(\omega) = 0 \quad \text{for all} \quad \omega \in \Omega_0.$$

$$\lambda_1(\omega) = \text{first entry time under } T \text{ of } \omega \text{ into } \Omega^1,$$

with $\lambda_1(\omega) = 0$ if $\omega \in \Omega^1$. In general

$$\lambda_k(\omega) = \text{first entry time under } T \text{ of } \omega \text{ into } \Omega^k,$$

with $\lambda_k(\omega) = 0$ if $\omega \in \Omega^k$.

The sequence γ_k, $k = 0, 1, 2, 3, \ldots$, of independent integer valued random variables is defined as follows:

$$\gamma_0(\omega) = \lambda_0(\omega) = 0 \quad \text{for all} \quad \omega \in \Omega_0,$$
$$\gamma_k(\omega) = \lambda_k(\omega) - \lambda_{k-1}(\omega), \quad k = 1, 2, 3, \ldots.$$

We have
$$\gamma_k(\omega) = \text{ first entry time of } T^{\lambda_{k-1}(\omega)}(\omega) \text{ into } \Omega^k, \tag{1}$$
$$\lambda_k(\omega) = \gamma_0(\omega) + \cdots + \gamma_k(\omega).$$

Note that $T^{\lambda_{k-1}(\omega)}(\omega) \in \Omega^{k-1}$, whence (1) shows that $\gamma_k(\omega)$ is constant on each piece of the partition $\mathcal{P}_k$; thus $\gamma_0, \gamma_1, \gamma_2, \ldots$ form a sequence of independent random variables; γ_k assumes the value 0 on $Q_{m_k}^k$. Further let us write

$$\gamma_{k,i} = \text{ value of } \gamma_k \text{ on } Q_{m_k-i}^k, \ 1 \le i < m_k.$$

15.45. The values $0, \gamma_{k,1}, \ldots, \gamma_{k,m_k-1}$ assumed by γ_k are related in a natural and useful manner to the values $0, R_{1,k}, R_{2,k}, \ldots, R_{m_k-1,k}, \ k = 1,2,3,\ldots$, which occur in the expression for the maximal spectral type of a rank one automorphism described in **15.27**. We have

$$\gamma_{k,i}(T) = R_{i,k}(T^{-1}), \gamma_{k,i}(T^{-1}) = R_{i,k}(T).$$

To see this one notes that the inverse of the rank one automorphism T is also a rank one automorphism obtained by cutting and stacking and one has a construction of T^{-1} in which Ω^k, Ω_k are respectively the base and the top of the kth stack for T^{-1}.

15.46. For $\omega \in \Omega_0$ let $l(\omega)$ be the last integer p for which $\omega \in \Omega^p$, i.e. $l(\omega) = p$, where p is given by

$$\lambda_0(\omega) = \lambda_1(\omega) = \cdots = \lambda_p(\omega) = 0, \lambda_{p+1}(\omega) \ne 0.$$

Let $f(\omega)$ equal the first re-entry time of ω into Ω_0:

$$f(\omega) = (\text{number of spacers above } \omega) + 1.$$

(Recall that τ denotes the automorphism induced on Ω_0 by T.) Then

$$\gamma_k(\omega) = 0, \quad \text{for } 1 \le k \le l(\omega),$$
$$\gamma_k(\omega) = \lambda_k(\tau(\omega)) + f(\omega), \quad k = l(\omega) + 1,$$
$$\gamma_k(\omega) = \gamma_k(\tau(\omega)), \quad k > l(\omega) + 1.$$

We therefore have in view of (1):

$$\sum_{p=1}^{\infty} (\gamma_p(\omega) - \gamma_p(\tau(\omega))) = f(\omega) + \lambda_{l(\omega)+1}(\tau(\omega)) - \sum_{p=1}^{l(\omega)+1} \gamma_p(\tau(\omega))$$

$$= f(\omega) = \quad (\text{number of spacers above } \omega) + 1. \tag{2}$$

15.47. Let Σ_k denote the group of permutations on $\{0, 1, \ldots, m_k - 1\}$ and Σ the restricted direct product of the Σ_k acting on

$$\Omega = \prod_{k=1}^{\infty} \{0, 1, \ldots, m_k - 1\}$$

by changing finitely many co-ordinates. We may view Σ as acting on Ω_0. Then the orbits of Σ and τ agree except on a countable subset of Ω_0. Note that if $\sigma \in \Sigma$, $\sigma = (\sigma_1, \ldots, \sigma_k, e, e, \ldots)$, then for each $n > k$, σ leaves invariant each element of $\mathcal{P}_n$. (Here e denotes the identity permutation on $\{0, 1, \ldots, m_k - 1\}$ for all k.) In particular, since each γ_n is $\mathcal{P}_n$ measurable, $\gamma_n \circ \sigma = \gamma_n$ for all $n > k$.

The Eigenvalue Group: Osikawa Criterion

15.48. Let $e(T)$ denote the group of eigenvalues of T and let f be as in **15.46**. The proposition and Theorem **15.50** below are essentially due to M. Osikawa [23].

15.49 Proposition. *Let $s \in [0, 1)$. Then $e^{2\pi i s} \in e(T)$ if and only if there exists a measurable function $\phi : \Omega_0 \to [0, 1)$ such that*

$$\phi(\tau(\omega)) = \phi(\omega) + sf(\omega) \pmod 1. \tag{3}$$

Proof. If a function ϕ satisfying (3) exists then $e^{2\pi i \phi}$ can be extended from Ω_0 to all of X in a natural way so that the extended function is an eigenfunction with eigenvalue $e^{2\pi i s}$: indeed if $x \in X$ is the pth spacer above ω, so that $x = T^p(\omega)$, define $\phi(x)$ by

$$\phi(x) = \phi(\omega) + ps \pmod 1.$$

The function $e^{2\pi i \phi}$, where ϕ is the extended function, is then an eigenfunction with eigenvalue $e^{2\pi i s}$.

On the other hand if $e^{2\pi i s}$ is an eigenvalue with eigenfunction ψ of absolute value one, then $\psi = e^{2\pi i \phi_1}$ for some measurable function ϕ_1 defined on X with $0 \leq \phi_1 < 1$. Set $\phi = \phi_1 |_{\Omega_0}$, then ϕ satisfies

$$\phi(\tau(\omega)) = \phi(\omega) + sf(\omega) \pmod 1,$$

which completes the proof of the proposition.

Let μ denote the Lebesgue measure on $\Omega_0 = [0, 1)$.

15.50 Theorem. *Let $s \in [0, 1)$, then $e^{2\pi i s} \in e(T)$ if and only if there exist real constants $c_n, n = 1, 2, \ldots$, such that*

$$\sum_{k=1}^{\infty} (s\gamma_k(\omega) - c_k) \tag{4}$$

converges (mod 1) for μ a.e. ω.

Proof. Suppose for an $s \in [0, 1)$, the series (4) converges (mod 1) μ a.e. to a function ϕ. Then (mod 1), for μ a.e. ω,

$$\phi(\tau(\omega)) - \phi(\omega) = \sum_{k=1}^{\infty} s(\gamma_k(\tau(\omega)) - \gamma_k(\omega)) = -sf(\omega) = (1-s)f(\omega),$$

by (2). By the proposition above we see that $e^{-2\pi is}$ is an eigenvalue of T. Since $e(T)$ is a group, $e^{2\pi is}$ is also an eigenvalue of T whenever (4) holds.

Conversely if $e^{-2\pi is} \in e(T)$ then by the proposition and (2) there exists $\phi : \Omega_0 \to [0, 1)$ such that (mod 1),

$$\phi(\tau^\nu(\omega)) - \phi(\omega) = \sum_{k=1}^{\infty} (1-s)(\gamma_k(\tau^\nu \omega) - \gamma_k(\omega)),$$

for all $\nu \in \mathbf{Z}$. If $\sigma = (\sigma_1, \sigma_2, \ldots, \sigma_n, e, e, \ldots) \in \Sigma$, then $\sigma(\omega) = \tau^{\nu(\omega)}(\omega)$ for some measurable function ν. Hence we have:

$$\phi(\sigma(\omega)) - \phi(\omega) = \sum_{k=1}^{\infty} (1-s)(\gamma_k(\sigma(\omega)) - \gamma_k(\omega))$$

$$= \sum_{k=1}^{n} (1-s)(\gamma_k(\sigma(\omega)) - \gamma_k(\omega)) \pmod 1,$$

since $\gamma_k(\sigma(\omega)) = \gamma(\omega)$ for $k > n$. (Recall that γ_k is $\mathcal{P}_k$ measurable.) Define

$$\phi_n(\omega) = \sum_{k=1}^{n} (1-s)\gamma_k(\omega),$$

and note that ϕ_n is $\mathcal{P}_1 \vee \mathcal{P}_2 \vee \cdots \vee \mathcal{P}_n$ measurable. The function $\psi_n = \phi - \phi_n$ satisfies

$$(\phi - \phi_n)(\omega) = \phi(\omega) - \sum_{k=1}^{n} (1-s)\gamma_k(\omega) \pmod 1,$$

which is invariant under all $\sigma = (\sigma_1, \ldots, \sigma_n, e, e, \ldots)$ and therefore measurable $\bigvee_{k=n+1}^{\infty} \mathcal{P}_k$.

Now $\phi = \phi_n + \psi_n$ and

$$e^{2\pi i\phi_n} \, \mathrm{E}\,(e^{2\pi i\psi_n}) = \mathrm{E}\,(e^{2\pi i\phi} \mid \mathcal{P}_1 \vee \cdots \vee \mathcal{P}_n) \to e^{2\pi i\phi} \quad \text{a.e.}$$

as $n \to \infty$. (Here E denotes the expectation or the conditional expectation.) Clearly there exist real constants A_n such that $\phi_n - A_n \to \phi \pmod 1$, indeed we can take

$$A_n = -Arg E(e^{2\pi i\psi_n}).$$

If we set $A_0 = 0$ and $c_k = A_k - A_{k-1}, k = 1, 2, \ldots$, then it follows that (mod 1),

$$\phi_n(\omega) - A_n = \sum_{k=1}^{n} ((1-s)\gamma_k(\omega) - c_k) \to \phi \text{ a.e. } [\mu].$$

This proves the theorem.

Restatement of Theorem 15.50

15.51. For any real number a let $[a]$ denote the largest integer $\leq a$, $\{a\} = a - [a]$ and

$$\langle a \rangle = \{a\} \text{ if } 0 \leq \{a\} \leq 1/2, \langle a \rangle = \{a\} - 1 \text{ if } 1/2 < \{a\} < 1.$$

We note that $|\langle a \rangle| \leq 1/2$ so that $\sum_{k=1}^{\infty} a_n$ converges (mod 1) if and only if $\sum_{k=1}^{\infty} \langle a_n \rangle$ converges. Using these remarks we can restate Theorem **15.50** in the following form.

For $s \in [0,1)$, $e^{2\pi i s} \in e(T)$ if and only if there exist real constants $c_k, k = 1, 2, \ldots$, such that any one of the following series converges mod 1 a.e. $[\mu]$,

(a)
$$\sum_{k=1}^{\infty} (\{s\gamma_k\} - c_k),$$

(b)
$$\sum_{k=1}^{\infty} (\langle s\gamma_k(\omega) \rangle - c_k),$$

(c)
$$\sum_{k=1}^{\infty} (\langle s\gamma_k(\omega) - c_k \rangle).$$

We can replace s by $-s$ or $1 - s$ in any of (a),(b),(c) above since eigenvalues form a group.

The Eigenvalue Group: Structural Criterion

15.52. We now give a criterion for $e^{2\pi i s}$ to be an eigenvalue of T in terms of the quantities $\gamma_{k,j}, 0 \leq j \leq m_k - 1, k = 1, 2, 3, \ldots$, which determine the rank one automorphism T. We need Theorem **15.54** below which is an analog for the circle group of a similar theorem for the real line. (See J. L. Doob [11], p 115, Theorem 2.7.) Recall that an infinite product $\prod_{k=1}^{\infty} a_k$ of complex numbers is said to be convergent if there is an M such that $\prod_{k=M}^{N} a_k$ converges to a non-zero complex number as N tends to infinity, which in turn holds true if and only if $\prod_{k=M}^{N} a_k$ tends to one as M, N tend to infinity. In case $0 \leq a_k \leq 1$, the non-convergence of the infinite product $\prod_{k=1}^{\infty} a_k$ is equivalent to the convergence to zero as N tends to infinity of the product $\prod_{k=M}^{N} a_k$ for every M.

15.53. Let Y be a random variable taking values in the circle group S^1. We will assume that our random variables are defined on a probability space $(W, \mathcal{C}, \mathcal{P})$. Let ν denote the distribution of Y and $\hat{\nu}$ its Fourier transform. Let $E(Y)$ and $\text{Var}(Y)$ denote respectively the expectation and variance of Y. We note that

$$E(Y^n) = \int_{S^1} z^n d\nu = \hat{\nu}(n), \quad n \in \mathbf{Z},$$

$$\text{Var}(Y) = \int_{S^1} |z - E(Y)|^2 d\nu = 1 - |E(Y)|^2 = 1 - |\hat{\nu}(1)|^2.$$

15.54 Theorem. *Let $Y_1, Y_2, Y_3, \ldots,$ be a sequence of independent S^1 valued random variables with distributions $\nu_1, \nu_2, \nu_3, \ldots,$ respectively. Then the following are equivalent:*

(a) *There exist real constants $c_k, k = 1, 2, 3, \ldots,$ such that if $Z_n = \prod_{k=1}^{n} Y_k e^{ic_k}$ then $Z_n, n = 1, 2, 3, \ldots,$ converges a.e. over a subsequence,*

(b) *for all integers $p \in \mathbf{Z}$, the infinite product*

$$\prod_{k=1}^{\infty} \mid \hat{\nu}_k(p) \mid^2$$

converges.

(c) *$\sum_{k=1}^{\infty} \mathrm{Var}(Y_k)$ converges,*

(d) *for some $p \neq 0$, the infinite product*

$$\prod_{k=1}^{\infty} \mid \hat{\nu}_k(p) \mid^2$$

converges.

Proof.   (a) implies (b). If $Z_{n_j}, j = 1, 2, 3, \ldots,$ converges a.e. then

$$Z_{n_l}(Z_{n_j})^{-1} = \prod_{k=n_j+1}^{n_l} Y_k e^{ic_k} \to 1$$

a.e. as $j, l \to \infty$, whence for all p, $\prod_{k=n_j+1}^{n_l} \hat{\nu}_k(p) e^{ipc_k} \to 1$ as $j, l \to \infty$. Therefore since $|\hat{\nu}_k(p)| \leq 1$, $\prod_{k=1}^{\infty} \mid \hat{\nu}_k(p) \mid^2$ is a convergent infinite product for all p.

Since $\mathrm{Var}(Y_k) = 1 - |\hat{\nu}_k(1)|^2$, it is easy to see that (b) implies (c) and that (c) implies (d).

We prove that (d) implies (a). Suppose that for some $p \neq 0$,

$$\prod_{k=1}^{\infty} \mid \hat{\nu}_k(p) \mid^2$$

is a convergent infinite product. Then

$$\prod_{k=j}^{l} \mid \hat{\nu}_k(p) \mid^2 \to 1$$

as $j, l \to \infty$. Since $\mid \hat{\nu}_k(q) \mid \leq 1$ the limit as $n \to \infty$ of $\prod_{k=l}^{n} \mid \hat{\nu}_k(q) \mid^2$ exists for each q and the resulting limit as a function of q is the Fourier transform of a probability measure, say ρ_ℓ. The functions $\hat{\rho}_\ell$ are non-decreasing and their limit as $\ell \to \infty$ is the Fourier transform of a probability measure, say ρ. Since $\hat{\rho}(p) = 1$ and $p \neq 0$ the measure ρ is the point mass at 1.

Let X_k be the random variable $X_k(x, y) = Y_k(x) \cdot \overline{Y_k}(y)$. (The bar denotes the complex conjugate.) Its distribution has Fourier transform $\mid \hat{\nu}_k(\cdot) \mid^2$. The finite products $\prod_{k=j}^{l} X_k$ converge in distribution to the point mass at 1 as $j, l \to \infty$. Hence they also converge in measure to the constant function 1. It follows that $\prod_{k=1}^{n} X_k, n = 1, 2, 3, \ldots$, converges a.e. over an increasing subsequence $n_1, n_2, n_3, \ldots$, of natural numbers. By Fubini's theorem we see that for some y the products

$$\prod_{k=1}^{n_j} Y_k(x) \cdot \overline{Y_k}(y), \quad j = 1, 2, 3, \ldots,$$

converge for a.e. x as $j \to \infty$. If we write $Y_k(y) = e^{ic_k}$, (a) follows, completing the proof of the theorem.

15.55. We apply this theorem to the random variables $Y_k = e^{2\pi i s \gamma_k}$, $k = 1, 2, \ldots$, of theorem **15.50**. Note that, in this case, if the products $\prod_{k=1}^{n} Y_k \cdot e^{ic_k}$, $k = 1, 2, 3, \ldots$, converge a.e. over a subsequence then the argument used in the proof of theorem **15.50** shows that the resulting limit extends to an eigenfunction of T with eigenvalue $e^{2\pi i s}$. Hence by theorem **15.50** the same product converges a.e. over the full sequence of natural numbers, possibly for some different constants c_k. Also note that

$$E(Y_k) = \frac{1}{m_k} \sum_{j=0}^{m_k-1} e^{2\pi i s \gamma_{k,j}},$$

$$\text{Var}(Y_k) = 1 - \frac{1}{m_k^2} \left| \sum_{j=0}^{m_k-1} e^{2\pi i s \gamma_{k,j}} \right|^2.$$

In view of Theorem **15.50** above we have at once the following characterization of the group $e(T)$. Write

$$\tilde{P}_k(z) = \sum_{j=0}^{m_k-1} z^{-\gamma_{k,j}}.$$

15.56 Theorem. *For $s \in [0, 1)$, the following are equivalent:*

(a) $e^{2\pi i s} \in e(T)$;
(b) the infinite product

$$\prod_{k=1}^{\infty} \frac{1}{m_k^2} \mid \tilde{P}_k(e^{2\pi i s}) \mid^2$$

is convergent;
(c) $\sum_{k=1}^{\infty} \text{Var}(e^{2\pi i s \gamma_k}) = \sum_{k=1}^{\infty} (1 - \frac{1}{m_k^2} \mid \tilde{P}_k(e^{2\pi i s}) \mid^2)$ is finite.

15.57 Corollary. *If either of the series*

$$\sum_{k=1}^{\infty} \left(\frac{1}{m_k} \sum_{j=0}^{m_k-1} |1 - e^{2\pi i s \gamma_{k,j}}| \right) \quad or \quad \sum_{k=1}^{\infty} \left(\frac{1}{m_k} \sum_{j=0}^{m_k-1} |1 - e^{2\pi i s \gamma_{k,j}}|^2 \right)$$

is finite then $e^{2\pi i s} \in e(T)$.

Proof. If the first series converges, then so does the second. We have

$$1 - \frac{1}{m_k^2} \left| \sum_{j=0}^{m_k-1} e^{2\pi i s \gamma_{k,j}} \right|^2 = \frac{1}{m_k^2} \sum_{j=0}^{m_k-1} \sum_{\ell=0}^{m_k-1} \left(1 - e^{2\pi i s \gamma_{k,j}} e^{-2\pi i s \gamma_{k,\ell}} \right)$$

$$= \frac{1}{m_k^2} \sum_{j<\ell} |e^{2\pi i s \gamma_{k,j}} - e^{2\pi i s \gamma_{k,\ell}}|^2$$

$$= \frac{1}{m_k^2} \sum_{j<\ell} |(1 - e^{2\pi i s \gamma_{k,j}}) - (1 - e^{2\pi i s \gamma_{k,\ell}})|^2$$

$$\leq \frac{2}{m_k^2} \sum_{j<\ell} \left(|1 - e^{2\pi i s \gamma_{k,j}}|^2 + |1 - e^{2\pi i s \gamma_{k,\ell}}|^2 \right)$$

$$= \frac{2(m_k - 1)}{m_k^2} \sum_{j=0}^{m_k-1} |1 - e^{2\pi i s \gamma_{k,j}}|^2.$$

Thus convergence of the second series implies condition (c) of Theorem **15.56.**, which proves the corollary.

15.58 Comments on Theorem 15.54. We note the close resemblance (already mentioned earlier) between the criterion for $e(T)$ obtained above and the expression for the maximal spectral type (up to discrete measures) obtained in section **15.29.** Since T and T^{-1} are spectrally equivalent, and as remarked in **15.45**, $R_{i,k}(T) = \gamma_{k,i}(T^{-1})$ and $R_{i,k}(T^{-1}) = \gamma_{k,i}(T)$, it follows that both sequences of polynomials: $P_k(z) = \sum_{j=0}^{m_k-1} z^{-R_{i,k}}$ and $\tilde{P}_k(z)$, give the eigenvalue group $e(T) = e(T^{-1})$. Thus $z \in e(T)$ if and only if $\prod_{k=1}^{\infty} \frac{1}{m_k^2} |P_k(z)|^2$ converges or equivalently if $\prod_{k=1}^{\infty} \frac{1}{m_k^2} |\tilde{P}_k(z)|^2$ converges. The maximal spectral type σ_0 of T or T^{-1} is given, up to a discrete measure, by either of the generalized Riesz products $\prod_{k=1}^{\infty} \frac{1}{m_k} |P_k(z)|^2$ or $\prod_{k=1}^{\infty} \frac{1}{m_k} |\tilde{P}_k(z)|^2$.

15.59 Theorem.
  (a) *If for $s \in [0,1), e^{2\pi i s} \in e(T)$, then the series $\sum_{k=1}^{\infty} \mathrm{Var}(|2\pi\langle s\gamma_k\rangle|)$ is convergent.*

  (b) *If the series $\sum_{k=1}^{\infty} \mathrm{Var}(2\pi\langle s\gamma_k\rangle)$ is convergent then $e^{2\pi i s} \in e(T)$.*

Proof. (a) Suppose $e^{2\pi i s} \in e(T), 0 \le s < 1$, then

$$1 - \frac{1}{m_k^2} \left| \sum_{j=0}^{m_k-1} e^{2\pi i s \gamma_{k,j}} \right|^2 \to 0$$

as $k \to \infty$. Without loss of generality we assume that

$$\left| \frac{1}{m_k} \sum_{j=0}^{m_k-1} e^{2\pi i s \gamma_{k,j}} \right| > 1/2.$$

For $z \ne 0$ write $z = |z| e^{i\theta}, -\pi \le \theta < \pi$. The map $\psi : z \to |\theta|$ is Lipschitz on any compact subset of the complex plane not containing the origin. Hence it is Lipschitz on $1/2 \le |z| \le 1$. Let C be the Lipschitz constant on this domain. Then

$$\left| \psi(e^{2\pi i s \gamma_k}) - \psi\left(\frac{1}{m_k} \sum_{j=0}^{m_k-1} e^{2\pi i s \gamma_{k,j}} \right) \right|^2.$$

$$\le C^2 \left| e^{2\pi i s \gamma_k} - \frac{1}{m_k} \sum_{j=0}^{m_k-1} e^{2\pi i s \gamma_{k,j}} \right|^2.$$

Since the variance of a random variable is smaller than the second moment around any other point,

$$\mathrm{Var}\left(\psi(e^{2\pi i s \gamma_k}) \right) = \mathrm{Var}(2\pi \, | \langle s \gamma_k \rangle |)$$

$$\le C^2 \, \mathrm{Var}(e^{2\pi i s \gamma_k}).$$

Thus (a) follows by Theorem **15.56**.

(b) The map $\phi(z) = e^{iz}$ is Lipschitz on any compact subset of the complex plane. Let C be Lipschitz constant for the domain $|z| \le 1$. We have

$$| e^{2\pi i s \gamma_k} - e^{(i \, \mathrm{E}(2\pi \langle s \gamma_k \rangle))} |$$

$$\le C \, | 2\pi \langle s \gamma_k \rangle - \mathrm{E}(2\pi \langle s \gamma_k \rangle) |.$$

Hence, by a similar argument as in (a), if the series $\sum_{k=1}^{\infty} \mathrm{Var}(2\pi \langle s \gamma_k \rangle)$ is finite then the series $\sum_{k=1}^{\infty} \mathrm{Var}(e^{2\pi i s \gamma_k})$ is finite and by Theorem **15.56** $e^{2\pi i s} \in e(T)$. This proves (b).

Remark. In case the m_k are bounded then it follows from a theorem of Y. Ito, T. Kamae and I. Shiokawa [16] that the converse of (b) holds, i.e., if $e^{2\pi i s} \in e(T)$ then $\sum_{k=1}^{\infty} \mathrm{Var}(2\pi \langle s \gamma_k \rangle)$ is finite.

15.60 Example. In the case of Chacon's automorphism, the height h_{k-1} of the $(k-1)^{th}$ stack is $h_{k-1} = \frac{3^k-1}{2}$, and γ_k assumes three values $0, 3^k, \frac{3^k+1}{2}$, with equal probability. The series

$$\sum_{k=1}^{\infty} \left(1 - \frac{1}{3^2} \left| 1 + e^{2\pi i s 3^k} + e^{2\pi i s \frac{3^k+1}{2}} \right|^2 \right).$$

can be shown to be divergent for all $s \neq 0$ so that Chacon's automorphism has no non-trivial eigenvalues. This proves the well known fact that Chacon's automorphism is weakly mixing.

An Expression for $\frac{d\sigma_\alpha}{d\sigma}$, $\alpha \in e(T)$

15.61. We first describe a very concrete necessary and sufficient condition for $e^{2\pi i s}$, $s \in [0,1)$ to be an eigenvalue of T. For each $k = 1, 2, 3 \ldots$, we define a function ψ_k on Ω_0 as follows: Let

$$q_k(\omega) = \quad \text{least integer} \geq 0 \text{ such that } T^{-q_k(\omega)}(\omega) \in \Omega_k$$

$$= h_k - \lambda_k(\omega) - 1.$$

If $\omega \notin \Omega^k$, $q_k(\tau\omega) = q_k(\omega) + f(\omega)$. Define

$$\psi_k(\omega) = e^{2\pi i s q_k(\omega)} = e^{2\pi i s(-\lambda_k(\omega)+h_k-1)}.$$

If $\lim_{n\to\infty} \psi_{k_n}(\omega)$ exists a.e. along some subsequence $k_n \to \infty$, then the limit function ψ satisfies $\psi(\tau\omega) = e^{2\pi i s f(\omega)} \psi(\omega)$, so that, by the proposition, $e^{2\pi i s} \in e(T)$. Conversely if $e^{2\pi i s} \in e(T)$ for some $s \in [0,1)$, then there exist real constants c_k such that $\sum_{k=1}^{\infty}(s\gamma_k(\omega) - c_k)$ converges a.e. (mod 1). Equivalently

$$\sum_{k=1}^{n}(s\gamma_k(\omega) - c_k) = s\lambda_n(\omega) - \sum_{k=1}^{n} c_k = s\lambda_n(\omega) - A_n$$

converges a.e. (mod 1), where $A_n = \sum_{k=1}^{n} c_k$. Since the A_n are constants, $s\lambda_k$ converges a.e.(mod 1) along a subsequence. For the same reason, since s, h_k are constants,

$$s q_k(\omega) = s h_k - s\lambda_k(\omega) - s$$

converges a.e. (mod 1) along a further subsequence, say k_n, to a function ϕ, so that $e^{2\pi i s q_{k_n}}$ converges a.e. to $e^{2\pi i \phi}$. We thus have:

15.62 Theorem. *For $s \in [0,1)$, $e^{2\pi i s} \in e(T)$ if and only if the sequence $\psi_k = e^{2\pi i s q_k}$, $k = 1, 2, 3, \ldots$, converges along a subsequence to a function ψ. This function ψ then extends in a natural way to an eigenfunction of T with eigenvalue $e^{2\pi i s}$.*

Note that our argument in fact shows that $e^{2\pi i s} \in e(T)$ if and only if given any increasing sequence k_n, $n = 1, 2, 3, \ldots$, of natural numbers there is a subsequence of it over which the functions $\psi_k, k = 1, 2, 3, \ldots$, converge a.e. to a function ψ which then extends to an eigenfunction of T with eigenvalue $e^{2\pi is}$. Any two such limits differ by a multiplicative constant of absolute value one. Note also that $e^{2\pi is} \in e(T)$ if and only if the ψ_k converge over a subsequence in the L^2 norm.

We note that the functions ψ_k vanish outside Ω_0. Since Ω_0 has finite measure, the ψ_k's are in $L^2(X, \mathcal{B}, m)$ with bounded L^2 norms. Any weak limit ψ of the collection $\{\psi_k : k = 1, 2, 3, \ldots\}$ satisfies the relation

$$\psi(\tau\omega) = e^{2\pi isf(\omega)}\psi(\omega).$$

If such a ψ is non-zero then it extends to an eigenfunction of T, and ψ is then an a.e. limit of the ψ_k's over a subsequence. Thus we see that either the ψ_k's converge weakly to zero or the ψ_k's converge a.e. over a subsequence to a function which extends to an eigenfunction with eigenvalue $e^{2\pi is}$.

15.63. The maximal spectral type σ of U_T is given (up to a discrete measure) by the weak limit as $n \to \infty$ of the measures

$$\prod_{k=1}^{n} \frac{1}{m_k} \mid P_k(z) \mid^2 dz.$$

If $\alpha \in S^1$, then the translate σ_α of σ by α is given by the weak limit of the measures $\prod_{k=1}^{n} \frac{1}{m_k} \mid P_k(\alpha z) \mid^2$. It is known that if $\alpha \in e(T)$ then σ_α and σ are mutually absolutely continuous.

Fix $s \in [0, 1)$, write $\alpha = e^{2\pi is}$ and let ψ_k be the functions as in Theorem **15.62** for this s. The correspondence $U_T^n 1_{\Omega_0} \leftrightarrow z^n, n \in \mathbf{Z}$ extends by linearity to an invertible isometry S from the closed linear span $\mathcal{H}$ of $\{U_T^n 1_{\Omega_0} : n \in \mathbf{Z}\}$ to $L^2(S^1, \sigma)$. We know from **15.27** that

$$1_{\Omega_0} = \left(\prod_{j=1}^{k} P_j(U_T) \right) 1_{\Omega_k},$$

and one sees similarly that

$$\psi_k = \left(\prod_{j=1}^{k} P_j(\overline{\alpha}U_T) \right) 1_{\Omega_k}, \quad S1_{\Omega_0} = \left(\prod_{j=1}^{k} P_j(\overline{z}) \right) S1_{\Omega_k},$$

$$S\psi_k = \left(\prod_{j=1}^{k} P_j(\overline{\alpha}\overline{z}) \right) S1_{\Omega_k}.$$

Since $S1_{\Omega_0} = 1$, we see that

$$S\psi_k = \prod_{j=1}^{k} \frac{P_j(\overline{\alpha}\overline{z})}{P_j(\overline{z})}.$$

By Theorem **15.62**, $\alpha \in e(T)$ if and only if the ψ_k's converge over a subsequence to a function ψ in the L^2 norm. Hence $\alpha \in e(T)$ if and only if $S\psi_k$'s converge over a subsequence in the L^2 norm. If ψ_k's converge over a subsequence in the L^2 norm to a function ψ, then $(S\psi_k)$'s will converge in the L^2 norm over the same subsequence to $S\psi$. Any two subsequential limits of the sequence $\psi_k, k = 1, 2, 3, \ldots$, differ by a constant of absolute value one, hence any two subsequential limits of the sequence $S\psi_k, k = 1, 2, 3, \ldots$, will also differ by a constant of absolute value one. In view of the remark after Theorem **15.62**, we see that if $\alpha \in e(T)$ then

$$\prod_{j=1}^{k} \left| \frac{P_j(\overline{\alpha}z)}{P_j(\overline{z})} \right|$$

converges in L^2 norm as $k \to \infty$ to the function $|S\psi|$, the convergence being over the full sequence of natural numbers. Hence, if $\alpha \in e(T)$ then

$$\prod_{j=1}^{k} \left| \frac{P_j(\overline{\alpha}z)}{P_j(\overline{z})} \right|^2$$

converges in $L^1(S^1, \sigma)$ to $|S\psi|^2$.

When $\alpha \in e(T)$, a subsequential limit ψ of the sequence $\psi_k, k = 1, 2, 3, \ldots$, is the restriction to Ω_0 of an eigenfunction ψ' with eigenvalue α. We have for such a subsequential limit ψ and $n \in \mathbf{Z}$;

$$(U_T^n \psi, \psi) = (U_T^n \psi' 1_{\Omega_0}, \psi' 1_{\Omega_0})$$
$$= (\alpha^n \psi' U_T^n 1_{\Omega_0}, \psi' 1_{\Omega_0})$$
$$= \alpha^n (U_T^n 1_{\Omega_0}, 1_{\Omega_0})$$
$$= \int_{S^1} (\alpha z)^n d\sigma$$
$$= \int_{S^1} z^n d\sigma_\alpha, \quad (\text{where } \sigma_\alpha(A) = \sigma(\alpha^{-1}A))$$
$$= \int_{S^1} z^n \frac{d\sigma_\alpha}{d\sigma} d\sigma.$$

But

$$(U_T^n \psi, \psi) = \int_{S^1} z^n |S\psi|^2 d\sigma, \quad n \in \mathbf{Z}.$$

Thus

$$\frac{d\sigma_\alpha}{d\sigma} = |S\psi|^2,$$

and we have proved:

15.64 Theorem. *If $\alpha \in e(T)$ then*

$$\frac{d\sigma_\alpha}{d\sigma} = \lim_{k \to \infty} \prod_{j=1}^{k} \left| \frac{P_j(\overline{\alpha}z)}{P_j(\overline{z})} \right|^2,$$

convergence being in the L^1 norm.

It is not known whether, when $\alpha \notin e(T)$, the measures σ and σ_α are mutually singular and further if

$$\lim_{k \to \infty} \prod_{j=1}^{k} \left| \frac{P_j(\overline{\alpha}z)}{P_j(\overline{z})} \right|^2 = 0 \quad \text{a.e.} \quad [\sigma]$$

in this case.

15.65 Asides. Let μ be the maximal spectral type of a measure preserving automorphism T. Call T spectrally prime if no measure strictly absolutely continuous with respect to μ can be the maximal spectral type of a measure preserving automorphism. A cyclic permutation on $\{1, 2, \ldots, p\}$, p a prime, is spectrally prime. It does not seem to be known if there exist other spectrally prime automorphism and whether prime automorphisms are spectrally prime.

Chapter 16

Additional Topics

In this chapter we treat some additional interesting topics. We first prove a theorem due to V. M. Alexeyev showing the existence of a bounded function with maximal spectral type.

Bounded Functions with Maximal Spectral Type

16.1 Theorem. *Let E be a spectral measure on a Borel space $(X, \mathcal{B})$ acting in a complex separable Hilbert space $\mathcal{H}$ which is the L^2 of a σ-finite measure space $(\Omega, \mathcal{A}, m)$. Then there exists a bounded function f in $\mathcal{H}$ such that the measure $\mu_f \colon \mu_f(A) = (E(A)f, f), A \in \mathcal{B}$ is the maximal spectral type of E.*

Recall the Hahn-Hellinger theorem in its first form. Let μ be the maximal spectral type of E. Then there exists sets $X = M_1 \supseteq M_2 \supseteq \cdots$, unique up to μ-null sets, and an invertible isometry

$$S : \mathcal{H} \leftrightarrow \sum_{n=1}^{\infty} L^2(X, \mathcal{B}, \mu\,|_{M_n})$$

such that for all $f \in \sum_{n=1}^{\infty} L^2(X, \mathcal{B}, \mu\,|_{M_n})$,

$$SE(A)S^{-1}f = 1_A f.$$

We write $Sf = (\tilde{f}_n)_{n=1}^{\infty}$, where $\tilde{f}_n \in L^2(X, \mathcal{B}, \mu\,|_{M_n})$ and denote by A_f the set $\cup_{n=1}^{\infty}\{x \in M_n : \tilde{f}_n(x) \neq 0\}$. Clearly μ_f is absolutely continuous with respect to μ_g if and only if $\mu_f(A_f - A_g) = 0$.

We will now show that given $f \in L^2(\Omega, \mathcal{A}, m)$, there exists a bounded function $g \in L^2(\Omega, \mathcal{A}, m)$ such that $\mu(A_f - A_g) = 0$. So let f be given and write

$$f^{(a,b)} = f(x) \text{ if } a \leq |\,f(x)\,| < b, \ f(x) = 0 \text{ otherwise.}$$

If $K > 1$, then

$$f^{(l)} =_{def} f^{(0,K^l)} \to f \quad \text{in } L^2(\Omega, \mu) \text{ as } l \to \infty.$$

If $Sf^{(l)} = (\tilde{f}_n^{(l)})_{n=1}^\infty$ and $Sf = (\tilde{f}_n)_{n=1}^\infty$, then for each n

$$\int_{M_n} \mid \tilde{f}_n(x) - \tilde{f}_n^{(l)} \mid^2 d\mu \to 0 \text{ as } l \to \infty.$$

Since convergence in L^2 implies convergence *a.e.* over a subsequence, by the diagonal method we can construct a sequence $(l_p)_{p=0}^\infty$, $l_0 = 1$ and sets $N_n \subset M_n$ such that

(a) $\mu(M_n - N_n) = 0$,

(b) $\forall x \in N_n, \tilde{f}_n^{(l_p)}(x) \to \tilde{f}_n(x)$ as $p \to \infty$.

Define

$$f(z, \cdot) = f^{(0,K)}(\cdot) + \sum_{p=1}^\infty z^{l_p} f^{(K^{l_{p-1}}, K^{l_p})}(\cdot).$$

The orthogonality of the functions $f^{(K^{l_{p-1}}, K^{l_p})}(\cdot)$, $p = 1, 2, \ldots$, implies that the series converges in L^2. Moreover,

$$
\begin{aligned}
f(z, \omega) &= f(\omega) \quad \text{if } 0 \leq \mid f(\omega) \mid < K \\
&= z^{l_1} f(\omega) \quad \text{if } K \leq \mid f(\omega) \mid < K^{l_1} \\
&\vdots \\
&= z^{l_p} f(\omega) \quad \text{if } K^{l_{p-1}} \leq \mid f(\omega) \mid < K^{l_p} \\
&\vdots
\end{aligned}
$$

from which we conclude that if $\mid z \mid < 1/K$, then $\mid f(z, \omega) \mid \leq \max(K, 1)$.

We are now going to investigate the spectral type of the functions $f(z, \omega)$. Let

$$Sf(z, \cdot) = (\tilde{f}_n(z, \cdot))_{n=1}^\infty$$

and $A_{n,z} = \{x : \tilde{f}_n(z, x) \neq 0\}$. We have

$$\tilde{f}_n(z, x) = \tilde{f}_n^{(0)}(x) + \sum_{p=1}^\infty z^{l_1}[\tilde{f}_n^{(l_p)}(x) - \tilde{f}_n^{(l_{p-1})}(x)], \tag{1}$$

the series converging in $L^2(X, \mathcal{B}, \mu)$. For $z = 1$, and $x \in N_n$, the series converges in the usual sense in view of (b) above. Hence by Abel's convergence theorem we conclude that $\tilde{f}_n(z, x)$ is analytic in the open disk $\mid z \mid < 1$. Hence for each $x \in N_n$, one of two possibilities hold. Either $\tilde{f}_n(z, x) \equiv 0$, or this function has at most a countable number of zeros in the disk $\mid z \mid< 1$. Let us denote by B_n the set of all $x \in N_n$ for which the second possibility holds. Since the identity $\tilde{f}_n(z, x) \equiv 0$ implies that all the coefficients of the series (1) vanish, which means $\tilde{f}_n^{l_p}(x) = 0$ for all l_p and $x \notin B_n$, so that

$$\tilde{f}_n(1, x) = \tilde{f}_n(x) = 0 \ \forall \ x \in N_n - B_n.$$

This means that $A_{n,1} \subseteq B_n$. Thus for all $x \in A_{n,1}$ the second possibility holds. Consider now the cartesian product $D \times A_{n,1}$ of the open unit disk D equipped with the Lebesgue measure ν and $A_{n,1}$ equipped with the measure μ_f. In this product the set $\{(z, x) : \tilde{f}_n(z, x) = 0\}$ has $\nu \times \mu_f$ measure zero since every x section of the set has m measure zero. In fact such sections consist of at most countably many points. Therefore, for almost every z, we have $\tilde{f}_n(z, x) \neq 0$ for a.e. $x \in A_{n,1}$. This implies that for a.e. z,

$$\mu_f(A_{n,1} - A_{n,z}) = 0.$$

If we choose a $z_0, \mid z_0 \mid< 1/K$, such that $\mu(A_{n,1} - A_{n,z_0}) = 0$ holds for all n and use the facts that $A_{n,z_0} \subseteq N_n \subseteq M_n$ and $\mu(M_n - N_n) = 0$, we obtain

$$\mu(A_f - A_{f(z_0,\cdot)}) = \mu(\cup_{n=1}^{\infty} A_{n,1} - \cup_{n=1}^{\infty} A_{n,z_0}) = 0.$$

This in turn implies that $\mu_f \ll \mu_{f(1,\cdot)} \ll \mu_{f(z_0,\cdot)}$. Since $f(z_0, \cdot)$ is bounded by $\max\{K, 1\}$, the theorem follows.

This theorem was originally published in *Vestnik. Mosc. Univ. Mat. Mekh. Asrton. Fis. Khim. No 5 (1958) 13-15*. It was translated from the Russian by A. Katok, see *Ergodic Theory and Dynamical System*, (1982) **2**, 259-261. The method yields more than what we have proved above as shown by K. Fraczek (see [4]).

A Result on Mixing

We will now prove a theorem due to D. Ornstein [5] which shows that a condition weaker than mixing implies mixing.

16.2 Theorem. *If T is a measure preserving automorphism on the unit interval $(0, 1)$ such that (a) every power of T is ergodic and (b) there exists a constant $K > 0$ such that*

$$\limsup_{n \to \infty} m(T^n A \cap B) < Km(A) \cdot m(B)$$

for all measurable A and B, then T is mixing.

The theorem is motivated by the following conjecture of Kakutani: If there is a constant $K > 0$ such that $\liminf_{n \to \infty} m(T^n A \cap B) > Km(A) \cdot m(B)$, then

T is mixing. This conjecture was proved to be false by N. Friedman and D. Ornstein See [5].

If every power of T is ergodic and $\alpha \neq 1$ is an eigenvalue of T then $\alpha^n, n \in \mathbb{Z}$, are all distinct. Indeed if ϕ is an eigenfunction of T with eigenvalue α then $\phi \circ T^n = \alpha^n \phi$. If $\alpha^n = 1$ for some $n \neq 0$, then ϕ would be a non-trivial function invariant under T^n, contradicting the ergodicity of T^n.

If every power of T is ergodic and ϕ is a non-trivial eigenfunction of T, then for all c, $\{x : \phi(x) = c\}$ has Lebesgue measure zero. To see this note that if the set $A_c = \{x : \phi(x) = c\}$ has positive measure, then by the Poincaré recurrence Lemma, for a.e. $x \in A_c$, $T^n x \in A_c$ for some n; on the other hand

$$c = \phi(T^n x) = \alpha^n \phi(x) = \alpha^n c \neq c,$$

(since $\alpha^n \neq 1$ for any n). Hence A_c has measure zero.

If every power of T is ergodic and if for some $K > 0$ and for all measurable A and B,

$$\limsup_{n \to \infty} m(A \cap T^n B) < K m(A) \cdot M(B),$$

then T is weakly mixing. This is weaker than saying that T is mixing, but it is nevertheless an intermediate step. If T is not weakly mixing then T admits an eigenvalue $\alpha \neq 1$. Let ϕ be an eigenfunction for α, $| \phi |= 1$. Let $z \in S^1$ be such that for every arc C containing z, $\phi^{-1}(C)$ has positive Lebesgue measure. Since every power of T is ergodic, in view of the discussion above, $m(\phi^{-1}(C))$ can be made as small as we please by choosing C small enough. Let $n_k \to \infty$ and be such that $\alpha^{n_k} \to 1$ Then

$$m((T^{-n_k} \phi^{-1}(C)) \cap \phi^{-1}(C)) = m(\phi^{-1}(\alpha^{-n_k} C \cap C)) \to m(\phi^{-1}(C))$$

as $k \to \infty$. At the same time

$$m((T^{-n_k} \phi^{-1}(C)) \cap \phi^{-1}(C)) \leq K(m(\phi^{-1}(C))^2.$$

If $m(\phi^{-1}(C))$ is chosen small enough, we arrive at a contradiction, proving that T is weakly mixing.

We can pick a sequence of integers (n_i) such that if C and D are intervals with rational end points, then

$$\lim_{i \to \infty} m((T^{n_i} C) \cap D) \text{exists.} \tag{1}$$

This follows from a standard diagonal procedure since there are only countable number of such C and D.

If T is not mixing, then the sequence (n_i) can be chosen so that, in addition to satisfying (1) , there is one pair of intervals with rational end points C_1, D_1 and

$$\lim_{i \to \infty} m[(T^{n_i} C_1) \cap D_1)] \neq m(C_1) \cdot m(D_1).$$

There is a measure u on $(0, 1) \times (0, 1)$ such that u is absolutely continuous with respect to the Lebesgue measure on $(0, 1) \times (0, 1)$ and if C and D are intervals with rational end points, then

$$u(C \times D) = \lim_{i \to \infty} m[(T^{n_i} C) \cap D].$$

To see this we note first that u is finitely additive on the class of rectangles with rational end points in the sense that if such a rectangle R is a finite pairwise disjoint union of rectangles $R_1, R_2, \ldots, R_n$ with rational end points, then $u(R) = \sum_{i=1}^{n} u(R_i)$. Therefore u extends to a finitely additive measure on the field generated by the class of rectangles with rational end points. By hypothesis (b) of the theorem the extended u is countably additive and absolutely continuous with respect to the Lebesgue measure on the unit square.

If A and B are two measurable sets in the unit interval then

$$u(A \times B) = \lim_{i \to \infty} m[(T^{n_i} A) \cap B]. \tag{2}$$

This holds if A and B are each a union of a finite number of intervals with rational end points. Let $A_n, B_n, n = 1, 2, \ldots$, be two sequences of such sets with $u(A \Delta A_n), u(B \Delta B_n) \to 0$ where $A \Delta A_n$ denotes the symmetric difference between A and A_n. Then

$$u(A \times B) = \lim_{n \to \infty} u(A_n \times B_n),$$

$$u(A_n \times B_n) = \lim_{i \to \infty} m[(T^{n_i} A_n) \cap B_n].$$

However, condition (b) implies that

$$\limsup_{i \to \infty} \mid m[(T^{n_i} A_n) \cap B_n] - m[(T^{n_i} A) \cap B] \mid$$

$$< K(m(A_n \Delta A)m(B) + m(B_n \Delta B)m(A) + m(B_n \Delta B)m(A_n \Delta A),$$

whence (2) follows.

The measure u is invariant under the automorphism $T \times T$ defined on the unit square by

$$T \times T(x, y) = (Tx, Ty).$$

i.e., u is a self joining of T. To see this we need only check this for sets of the form $A \times B$:

$$u(A \times B) = \lim_{i \to \infty} m[(T^{n_i} A \cap B],$$

$$u(TA \times TB) = \lim_{i \to \infty} m[(T^{n_i} TA) \cap TB] = \lim_{i \to \infty} m[T([T^{n_i} A] \cap B)]$$

Since T is weakly mixing $T \times T$ is ergodic (with respect to the Lebesgue measure on $(0, 1) \times (0, 1)$). Since u is absolutely continuous with respect to $m \times m$ and invariant under $T \times T$, it is a constant multiple of $m \times m$. Since u and $m \times m$ are probability measures $u = m \times m$. This gives a contradiction since $u(C_1 \times D_1) \neq m(C_1 \times D_1)$. The theorem follows.

A Result On Multiplicity

We now prove a theorem due to G. Goodson and M. Lemańczyk [1] concerning the multiplicity of the spectrum of a T which is conjugate to T^{-1}.

16.3 Theorem. *Let T be an ergodic measure preserving automorphism on a standard probability space (X, μ) and assume that there exists a measure preserving automorphism S such that $STS^{-1} = T^{-1}$. Then the essential values of the multiplicity function of U_T, restricted to the ortho-complement of the subspace*

$$\{f \in L^2(X, \mu) : f(S^2) = f\},$$

are even, ∞ being regarded as an even number.

This follows as an immediate consequence of the following:

16.4 Theorem. *Let $T : L^2(X, \mu) \to L^2(X, \mu)$ be a unitary operator which preserves real-valued functions and admits a unitary S, also preserving real-valued functions, which conjugates T and T^{-1}, i.e., satisfies $STS^{-1} = T^{-1}$. Then T preserves the ortho-complement C of the subspace*

$$\{f \in L^2(X, \mu) : S^2(f) = f\},$$

and on C the essential values of the multiplicity function of T are even (∞ is considered as an even number).

Proof. If $\mu_\infty, \mu_1, \mu_2, \ldots$, are the measures associated with T as per the second form of the Hahn-Hellinger theorem, then a cardinal n is called an essential value of T if μ_n is non-trivial. With this clarification we now proceed to the proof of the theorem.

Let E_S denote the spectral measure of S^2. The equality $ST = T^{-1}S$ implies that S^2 and T commute. Indeed, $T^{-1}S^{-1} = S^{-1}T$, and

$$S^2T = SST = ST^{-1}S = ST^{-1}S^{-1}S^2 = SS^{-1}TS^2 = TS^2.$$

For any integer n, and $f \in L^2(X, \mu)$

$$(S^{2n}Tf, Tf) = (S^{2n}f, f) = \int_{S^1} z^n d\sigma_f,$$

where $\sigma_f(\cdot) = (E_S(\cdot)f, f)$ is the spectral type of f. We see that f and Tf have the same spectral type with respect to S^2. Let S^+, S^- denote the parts of the unit circle in the open upper half plane and the open lower half plane respectively. Let supp σ mean the support of σ, not necessarily closed. Clearly, the subspaces

$$H_1 = \{f : S^2f = f\} = \{f : \text{supp } \sigma_f \subseteq \{1\}\},$$

$$H_{-1} = \{f : S^2f = -f\} = \{f : \text{supp} \sigma_f \subseteq \{-1\}\},$$

$$\mathcal{P}_1 = \{f : \text{supp } \sigma_f \subseteq S^+\}, \mathcal{P}_2 = \{f : \text{supp } \sigma_f \subseteq S^-\}$$

are mutually orthogonal, invariant under S^2 and T. Further

$$C = H_1^\perp = H_{-1} \oplus \mathcal{P}_1 \oplus \mathcal{P}_2 = H_{-1} \oplus \mathcal{K},$$

where $\mathcal{K}$ denotes the direct sum of $\mathcal{P}_1$ and $\mathcal{P}_2$. Since S^2 preserves real-valued functions in $L^2(X, \mu)$, it commutes with complex conjugation and its spectral measure is symmetric. For any $f \in L^2(X, \mu)$, $n \in \mathbb{Z}$,

$$(S^{2n}S\overline{f}, S\overline{f}) = (S^{2n}\overline{f}, \overline{f}) = \overline{(S^{2n}f, f)},$$

so that $\sigma_{S\overline{f}} = \tilde{\sigma}_f$, where $\tilde{\sigma}_f(A) = \sigma(A^{-1})$. So the map $f \to S\overline{f}$ is an invertible isometry between $\mathcal{P}_1$ and $\mathcal{P}_2$. Further, for $f \in \mathcal{P}_1$ and for all $n \in \mathbb{Z}$,

$$(T^n f, f) = (ST^n f, Sf) = (T^{-n} Sf, Sf)$$

$$= (Sf, T^n Sf) = \overline{(T^n Sf, Sf)} = (T^n S\overline{f}, S\overline{f}).$$

so that f and $S\overline{f}$ have the same spectral type with respect to T. Clearly the restriction of T to $\mathcal{K}$ has even essential values. We now show that the same holds for the restriction of T to H_{-1}. Indeed, as seen above f and $S\overline{f}$ have the same spectral type with respect to T. In addition if $f \in H_{-1}$, then we have $S\overline{f}$ orthogonal to $T^n f$ for all $n \in \mathbb{Z}$ since

$$(T^n f, S\overline{f}) = (ST^n f, S^2\overline{f}) = (T^{-n} Sf, -\overline{f}) = -(Sf, T^n \overline{f}) = -(T^n f, S\overline{f}).$$

If $f, g \in H_{-1}$ and g is orthogonal to $T^n f$ and $T^n S\overline{f}$ for all n, then $S\overline{g}$ is also orthogonal to $T^n f$ and $T^n S\overline{f}$ for all n since

$$(T^n f, S\overline{g}) = (ST^n f, S^2\overline{g}) = -(ST^n f, \overline{g}) = \overline{(T^{-n} S\overline{f}, g)} = 0.$$

These observations ensure that the essential multiplicity of T restricted to H_{-1} is also even and the theorem is proved.

Exercise 1. With the notation of theorem **16.3** show that if U_T has simple spectrum then $S^2 = I$. (See G. R. Goodson, A. del Junco, M. Lemańczyk, D. J. Rudolph [2]). In case S is weak mixing the essential values of the multiplicity function of U_T are even. (See G. R. Goodson and M. Lemańczyk [1].)

Exercise 2. If S^2 and T are ergodic and satisfy the conditions of **16.3**, then T and S are weakly mixing.

16.5. The rest of this chapter is devoted to the construction and proof of Ornstein's mixing rank one automorphisms [6]). This needs combinatorial and probabilistic preliminaries which will be covered in the next three sections.

Combinatorial and Probabilistic Lemmas

16.6. A Combinatorial Lemma. *Let* m, n *be positive integers,* $n < m$. *Consider the set* E *of* $m - n$ *pairs*

$$(1, n + 1), (2, 2 + n), (3, 3 + n), \cdots , (m - n, m).$$

We can divide E *into two sets* E_1 *and* E_2, *each containing at least* $[\frac{m-n}{4}]$ *pairs, such that no integer occurs in more than one pair of* E_1 *and the same holds for* E_2.

Proof. We use the division algorithm and write

$$m - n = d \cdot n + r, \quad 0 \leq r < n.$$

If $d = 0$, then among the $m - n$ pairs, each first co-ordinate is less than every second co-ordinate, so no integer is repeated more than once in E, therefore we can divide the $m - n$ pairs into two sets E_1 and E_2 one containing $[\frac{m-n}{2}] + 1$ ($\geq [\frac{m-n}{4}]$) pairs and the other containing the rest, (which is also $\geq [\frac{m-n}{4}]$).

If $d = 1$, then we distribute the first n pairs $(1, n + 1), (2, 2 + n), (3, 3 + n), \cdots , (n, 2n)$, into E_1 and E_2 so that each contains at least $[\frac{n}{2}]$ pairs. The remaining r pairs are $(n + 1, 2n + 1), (n + 2, 2n + 2), \cdots , (n + r, 2n + r)$. The first co-ordinates of these occur as the second co-ordinates in pairs $(1, n + 1), (2, 2 + n), \cdots , (r, n + r)$, which are already distributed. We can assign $(n + i, 2n + i)$ to E_1 if $(i, n + i) \in E_2$ and to E_2 otherwise, $1 \leq i \leq r$. This will ensure that E_1, E_2 will each contain at least $[\frac{m-n}{4}]$ pairs and no integer occurs in more than one pair of E_1 and the same holds for E_2.

If $d \geq 2$, we divide the $m - n$ pairs into d blocks of n consecutive pairs and a last block of r pairs. We assign alternately the first block to E_1 the second block to E_2 and so on until all the blocks are exhausted. It is easy to see that E_1 and E_2 will each contain more than $[\frac{m-n}{4}]$ elements. Further no integer occurs in more than one pair of E_1, and the same holds for E_2. Indeed, if an integer i occurs in more than one pair then these pairs are necessarily $(i - n, i)$ and $(i, i + n)$ which belong to consecutive blocks since their first co-ordinates differ by n. This proves the lemma.

The next lemma is a well known fact from the theory of large deviation.

16.7 Lemma. *Given a positive integer n, a sequence $(X_i)_{i=1}^n$ of bounded, independent, identically distributed random variables with mean m and distribution μ, and a real number $x > m$, there exists γ, $0 < \gamma < 1$, depending only on x and μ, such that*

$$P\{S_n > nx\} < \gamma^n,$$

where $S_n = \sum_{i=1}^n X_i$.

Proof. Let $s > 0$. Then

$$\frac{S_n}{n} - x > 0 \Leftrightarrow \exp s\left(\frac{S_n}{n} - x\right) > 1.$$

Further

$$E\left(e^{\frac{s}{n}S_n}\right) e^{-sx} \geq P\left(\exp s\left(\frac{S_n}{n} - x\right) > 1\right) = P\left(\frac{S_n}{n} > x\right). \tag{1}$$

But $(X_i)_{i=1}^n$ are independent, whence

$$\mathbb{E}[e^{\frac{s}{n}S_n}] = \prod_{i=1}^n \mathbb{E}[e^{\frac{s}{n}X_i}] = (\mathbb{E}[e^{\frac{s}{n}X_1}])^n$$

$$= \exp(n \log \mathbb{E}[e^{\frac{s}{n}X_1}])$$

Let us define Λ as follows:

$$\Lambda(t) = \log \mathbb{E}[e^{tX_1}],$$

so that (1) takes the form

$$\mathbb{P}\left\{\frac{S_n}{n} > x\right\} \leq \exp\left(-n\left(\frac{s}{n}x - \Lambda\left(\frac{s}{n}\right)\right)\right),$$

and this is true for any positive real number s, so for any positive real number γ we have

$$\mathbb{P}\left\{\frac{S_n}{n} > x\right\} \leq \exp(-n(\gamma x - \Lambda(\gamma))).$$

Define now Λ^* by:

$$\Lambda^*(x) = \sup_{\gamma \geq 0}(\gamma x - \Lambda(\gamma))$$

Now $\lim_{\gamma \to 0} \frac{\Lambda(\gamma)}{\gamma} = m$, so if $x > m$, then $\lim_{\gamma \to 0}(x - \frac{\Lambda(\gamma)}{\gamma}) > 0$, whence $\Lambda^*(x) > 0$ if $x > m$. The lemma follows.

This is needed in the proof of the crucial probabilistic arithmetical lemma proved below. In what follows $card(\cdot)$ means the cardinality of the set which appears in the parenthesis.

16.8 Lemma. *Let*

(a) *K be a positive integer,*
(b) *$X = \{i \in Z :| i |\leq \frac{K}{2}\}$, $\Omega_m = X^m$, $P_m = $ uniform distribution on Ω_m,*
(c) *$\omega = (\omega_1, \omega_2, \cdots, \omega_m)$ denote a point in Ω_m, $x_i, i = 1, 2, \ldots, m$ the co-ordinate random variables on Ω_m,*
(d) *$C_{n,l}(\omega) = card\{i : x_{i+n}(\omega) - x_i(\omega) = l\}$, where n is a positive integer and l is an integer.*

Then given $\alpha > 1$, $\epsilon > 0$ and a positive integer N, there exists an $m = m_0 > N$ such that,

$$P_{m_0}(\cap_{1 \leq n \leq (1-\epsilon)m_0} \cap_{l \in Z} \left\{\omega : C_{n,l} \leq \frac{\alpha}{K}(m_0 - n)\right\}) > 1 - \epsilon$$

Proof. Note that $| x_{i+n}(\omega) - x_i(\omega) |\leq K$, whence $C_{n,l} = 0$ if $l > K$. For any positive B the intersection $\cap_{l \in Z}\{\omega : C_{n,l}(\omega) < B\}$ is indeed the intersection taken over $| l |\leq K$.

Let $n < m$ and divide the $m - n$ pairs

$$(1, n + 1), (2, n + 2), \cdots, (m - n, m)$$

into disjoint sets E_1 and E_2, each of cardinality $\geq \frac{m-n}{4}$ and such that no integer occurs in more than one pair of E_1 and the same holds for E_2. The random variables $x_{i+n} - x_i$, $(i, i+n) \in E_1$ are then independent, and they are obviously identically distributed. Let $| l |\leq K$, and

$$F_i = \{\omega : x_{i+n}(\omega) - x_i(\omega) = l\}.$$

Note that

$$P_m(F_i) = \sum_{|a| \leq \frac{K}{2}} P_m(\{\omega : x_{i+n} = l - a, x_i(\omega) = a\}) < \frac{1}{K},$$

since the number of terms in the summation is at most $K+1$ and each summand is at most $\frac{1}{(K+1)^2}$.

The random variables 1_{F_i}, $(i, i+n) \in E_1$ are bounded, independent and identically distributed with the expected value of 1_{F_i} less than $\frac{1}{K}$. Write

$$D_1 = \sum_{(i,i+n) \in E_1} 1_{F_i}.$$

By the Lemma from the theory of large deviation proved above

$$P_m \left(\left\{ \omega : D_1(\omega) \geq \frac{\alpha}{K} \mid E_1 \mid \right\} \right) \leq \gamma^{|E_1|},$$

where $0 < \gamma < 1$, and $\mid E_1 \mid$ denotes the number of elements in E_1.

Similarly,

$$P_m \left(\left\{ \omega : D_2(\omega) \geq \frac{\alpha}{K} \mid E_2 \mid \right\} \right) \leq \gamma^{|E_2|},$$

where $D_2 = \sum_{(i,i+n) \in E_2} 1_{F_i}$. Observe now that $C_{n,l} = D_1 + D_2$, whence,

$$\left\{ \omega : C_{n,l}(w) \geq \frac{\alpha}{K}(m - n) \right\} \subseteq \cup_{j=1}^2 \left\{ \omega : D_j(\omega) \geq \frac{\alpha}{K} \mid E_j \mid \right\}.$$

Since the cardinalities of E_1 and E_2 are not smaller than $[\frac{m-n}{4}]$, we get

$$P_m \left(\left\{ \omega : C_{n,l}(\omega) \geq \frac{\alpha}{K}(m - n) \right\} \right) \leq 2\gamma^{[\frac{1}{4}(m-n)]}$$

Write

$$A = \cup_{1 \leq n \leq (1-\epsilon)m} \cup_{|l| \leq K} \left\{ \omega : C_{n,l}(\omega) > \frac{\alpha}{K}(m - n) \right\}.$$

Then,

$$P_m(A) \leq m \times (2K + 1) \times 2\gamma^{[\frac{1}{4}\epsilon m]} \to 0,$$

as $m \to \infty$, since $0 < \gamma < 1$. We can find $m = m_0 > N$ such that $P_{m_0}(A) \leq \epsilon$, whence $P_{m_0}(X - A) > 1 - \epsilon$. This proves the lemma.

Ornstein [6] stated this lemma in a purely arithmetical form as follows (with the probabilistic proof given above).

Given $\epsilon > 0$ and even positive integers N and K and $\alpha > 0$, we can find an $m > N$ and a sequence $\{a_i\}$, with $i = 1, \ldots, m$, of integers such that

(i) $\mid \sum_j^{j+k} a_i \mid \leq K$ for all $1 \leq j \leq j + k \leq m$.

(ii) Let $H(l, k)$ be the number of j such that $\sum_j^{j+k} a_i = l$ where $1 \leq j \leq j + k \leq m$. If $k \leq (1 - \epsilon)m$ then $H(l, k) < \alpha(K)^{-1}(m - k)$.

(Note that $H(l, k)$ in this statement is our $C(k, l)$, the order of the arguments being interchanged.)

Rank One Automorphisms by Construction

16.9. Using the cutting and stacking method we define a family of measure preserving automorphisms, called rank one automorphisms, as follows:

Let B_0 be the unit interval equipped with the Lebesgue measure. At the stage one we divide B_0 into p_0 equal parts, add spacers and form a stack of height h_1 in the usual fashion. At the kth stage we divide the stack obtained at $(k-1)^{th}$ stage into p_{k-1} equal columns add spacers and obtain a new stack of height h_k. If during the kth stage of our construction the number of spacers put above the jth column of the $(k-1)^{th}$ stack is $a_j^{(k-1)}$, $0 \le a_j^{(k-1)} < \infty$, $1 \le j \le p_{k-1}$, then we have

$$h_k = p_{k-1}h_{k-1} + \sum_{j=1}^{p_{k-1}} a_j^{(k-1)}$$

Proceeding thus we get a rank one automorphism T on a certain measure space $(X, \mathcal{B}, m)$, (m is the Lebesgue measure on X), which may be finite or σ-finite depending on the number of spacers added. For each $k = 1, 2, 3, \ldots$, let $_kJ_\alpha$ denote the αth level of the kth stack and when we divide the kth stack into columns we denote by $_kJ_\alpha^i$ the piece of $_kJ_\alpha$ in the ith column.

So the construction of a rank one automorphism needs two parameters, viz., the sequence $(p_k)_{k=0}^\infty$: the parameter of cutting, and $((a_j^{(k)})_{j=1}^{p_k})_{k=0}^\infty$: the parameter of spacers. We have by definition:

$$T =_{def} T_{(p_k,(a_j^{(k)})_{j=1}^{p_k})_{k=0}^\infty}$$

16.10 Lemma. *If for each $k \ge 1$, the integers $a_i^{(k-1)}$ are $\le 2h_{k-2}$ and if $p_{k-1} \ge (k-1)^2$, then the Lebesgue measure of the total space X remains bounded by the constant*

$$A = \prod_{j=1}^\infty \left(1 + \frac{2}{j^2}\right).$$

Proof. The Lebesgue measure of the union of intervals in the stack of height h_k is $c_k = \frac{h_k}{p_0 \cdot p_2 \cdots p_{k-1}}$. Since we add spacers at each stage of our construction, $(c_k)_{k=1}^\infty$ is a non-decreasing sequence. Since $a_j^{(k-1)} \le 2h_{k-2}$,

$$h_k = p_{k-1}h_{k-1} + \sum_{i=1}^{P(k-1)} a_i^{k-1} \le p_{k-1}(h_{k-1} + 2h_{k-2}),$$

$$c_k = \frac{h_k}{p_0 \cdot p_1 \cdots p_{k-1}} \le \frac{h_{k-1}}{p_0 \cdot p_1 \cdots p_{k-2}} + \frac{2h_{k-2}}{p_0 \cdot p_1 \cdots p_{k-2}},$$

$$c_k \le c_{k-1} + \frac{2}{p_{k-2}}c_{k-2} \le \left(1 + \frac{2}{p_{k-2}}\right)c_{k-1}.$$

On iterating we get, for $k \geq 3$,

$$c_k \leq c_0 \prod_{j=1}^{k-2} \left(1 + \frac{2}{p_j}\right) = \prod_{j=1}^{k-2} \left(1 + \frac{2}{p_j}\right)$$

since $c_0 = 1$.

Since $p_j \geq j^2$,

$$c_k \leq \prod_{j=1}^{\infty} \left(1 + \frac{2}{j^2}\right) = A,$$

and so $c =_{def} \lim_{k \to \infty} c_k$ (which is the Lebesgue measure of the total space X) is bounded by A. The Lemma is proved.

Let us normalise the Lebesgue measure on X and call the resulting measure ν. Let $0 < \epsilon \leq \frac{1}{p_0 \cdot p_1 \cdots p_{k-2} \cdot A}$. Then, since $m(X) \leq A$,

$$\epsilon p_k \cdot \nu(_{k+1}J_1) \leq \frac{1}{p_0 \cdot p_1 \cdots p_{k-2} \cdot A} \cdot \frac{p_k}{p_0 \cdot p_1 \cdots p_k \cdot m(X)}$$

$$\leq \nu(_{k-1}J_1) \cdot \nu(_kJ_1).$$

If in addition $4 \leq \epsilon p_k$, then

$$4\nu(_{k+1}J_1) \leq \nu(_{k-1}J_1) \cdot \nu(_kJ_1).$$

These observations will be needed later.

Ornstein's Class of Rank One Automorphisms

16.11. In Ornstein's construction, $a_i^{(k)}$, $1 \leq i \leq p_k - 1$, are chosen stochastically as follows: we choose independently, using the uniform distribution on the set $X_k = \{i : |i| \leq \frac{h_{k-1}}{2}\}$, the numbers $(x_{k,i})_{i=1}^{p_k-1}$. The integer $x_{k,0} = 0$ and the integer x_{k,p_k} is chosen deterministically in X_k.

We put, for $1 \leq i \leq p_k$,

$$a_i^{(k)} = h_{k-1} + x_{k,i} - x_{k,i-1}.$$

One sees that

$$h_{k+1} = p_k(h_k + h_{k-1}) + x_{k,p_k}.$$

So the deterministic sequences of positive integers $(p_k)_{k=0}^{\infty}$ and $(x_{k,p_k})_{k=0}^{\infty}$ completely determine the sequence of heights $(h_k)_{k=1}^{\infty}$. The total measure of the resulting measure space is less than A if $p_k > k^2$ for $k \geq 1$.

16.12. In Ornstein's construction the numbers x_{k,p_k} are chosen between 1 and 4 to ensure the ergodicity of each power T^n. However, as observed by El Houcein, this is not necessary as we shall see.

We thus have a probability space

$$\prod_{k=0}^{\infty} \Omega_k, \ \Omega_k = X_k^{p_k-1}$$

equipped with the product probability measure $\otimes_{k=1}^{\infty} P_k$, where P_k is the uniform probability on Ω_k, which indexes the family of Ornstein's automorphisms. We denote this space by $(\Omega, \mathcal{A}, P)$. So $x_{k,i}$ is a projection from Ω onto the ith co-ordinate space of Ω_k, $1 \leq i \leq p_k - 1$. Naturally each point $\omega = (\omega_k = (x_{k,i}(\omega))_{i=1}^{p_k-1})_{k=0}^{\infty}$ in Ω defines the spacers and the cutting parameters of a rank one automorphism which we denote by T_ω. We will show that if $p_k, k = 0, 1, 2, \ldots$ go to ∞ sufficiently fast then T_ω is mixing for a.e. $\omega \in \Omega$.

The following lemma is due to El Houcein. An automorphism is said to be totally ergodic if all its powers are ergodic. It is an easy exercise to see that a measure preserving automorphism is totally ergodic if and only if no root of unity is its eigenvalue.

16.13 Total Ergodicity Lemma. *If the cutting parameter $(p_k)_{k=1}^{\infty}$ is not bounded then the associated Ornstein's automorphism T_ω is totally ergodic for almost $\omega \in \Omega$.*

Proof. We need the following characterization of the eigenvalues of a rank one automorphism given in **15.56**.

If T is a rank one automorphism with parameters $(p_k, (a_i(k))_{i=1}^{p_k})_{k=0}^{\infty}$, and if

$$P_k = \sum_{j=0}^{p_k-1} z^{jh_k + \sum_{i=0}^{j} a_i^{(k)}}, \quad \left(a_0^{(k)} = 0\right).$$

then a complex number z is an eigenvalue of T if and only if

$$\sum_{k=1}^{\infty} \left(1 - \frac{1}{p_k^2} \mid P_k(z) \mid^2\right) < \infty.$$

We apply this criterion to the setting on hand. Here

$$P_k(z) = \sum_{j=0}^{p_k-1} z^{j(h_k + h_{k-1}) + x_{k,j}},$$

so, for $z \neq 1$,

$$\frac{1}{p_k^2} \mid P_k(z) \mid^2 = \frac{1}{p_k} + \frac{1}{p_k^2} \sum_{p \neq q} z^{(p-q)(h_k + h_{k-1})} z^{x_{k,p} - x_{k,q}}.$$

Integrating with respect to $\mathbb{P}$ we have:

$$\int_\Omega \frac{1}{p_k^2} \mid P_k(z) \mid^2 dP = \frac{1}{p_k} + \frac{1}{p_k^2} \sum_{p \neq q} z^{(p-q)(h_k + h_{k-1})} \frac{1}{(h_{k-1}+1)^2} \left| \sum_{|s| \leq \frac{h_{k-1}}{2}} z^s \right|^2 .$$

Now

$$\frac{1}{(h_{k-1}+1)^2} \left| \sum_{|s| \leq \frac{h_{k-1}}{2}} z^s \right|^2 \leq \frac{1}{(h_{k-1}+1)^2} \frac{\mid 1 - z^{h_{k-1}+1} \mid^2}{\mid 1 - z \mid^2}$$

$$\leq \frac{1}{h_{k-1}^2} \frac{4}{\mid 1 - z \mid^2} .$$

Thus

$$\int_\Omega \frac{1}{p_k^2} \mid P_k(z) \mid^2 dP \leq \left(\frac{1}{p_k} + \frac{4}{h_{k-1}^2 \mid 1 - z \mid^2} \right) \to 0$$

as $k \to \infty$ over any subsequence of natural numbers over which p_k diverges to ∞. So we can extract a subsequence which converges almost surely to 0:

$$\frac{1}{(p_{k_n})^2} \left| \sum_{j=0}^{p_{k_n}-1} z^{j(h_{k_n}+h_{k_n-1})+x_{k_n,j}} \right|^2 \to 0$$

$\mathbb{P}$ a.s. Thus for $z \neq 1$

$$\sum_{k=1}^{\infty} \left(1 - \frac{1}{p_k^2} \mid P_k(z) \mid^2 \right) = \infty$$

$\mathbb{P}$ a.s. So we have

$$\mathbb{P}\{\omega : z \text{ is an eigenvalue of } T_\omega\} = 0.$$

Hence

$$\cup_{n=1}^{\infty}\{\omega : e^{\frac{2\pi i}{n}} \text{ is an eigenvalue of } T_\omega\},$$

has P measure zero. The lemma follows.

Mixing Rank One Automorphisms

16.14. It is not known if, under the hypothesis of the above lemma, T_ω is weakly mixing for a.e. ω. The rest of this chapter will be devoted Ornstein's theorem, viz., that if the cutting parameters $p_k, k = 0, 1, 2, \dots$ go to infinity sufficiently fast then a.e. T_ω is mixing. The choice of p_k's is done as follows:

Suppose we have chosen $p_0, p_1, \cdots, p_{k-1}$ together with the associated spacers and obtained a stack of height h_k. We wish to choose p_k. We apply the probabilistic arithmetical lemma with $K = h_{k-1}$, $\alpha = \frac{5}{4}$, $0 < \epsilon =_{def} \epsilon_k \le \frac{1}{p_0 \cdot p_1 \cdots p_{k-2} \cdot A}$ and also less than 10^{-k-3}, $N = 10^k h_k$ and choose $m_0 > N$ such that $m_0 \epsilon_k > 4$ and $P_{m_0}(L_k) \ge 1 - \epsilon_k$, where

$$L_k = \cap_{1 \le n \le (1-\epsilon_k)m_0} \cap_{|l| \le h_{k-1}} \left\{ \omega : C_{n,l} \le \frac{\alpha}{K}(m_0 - n) \right\}$$

We let $p_k = m_0 + 1$. Proceeding thus we choose the cutting parameters $p_k, k = 0, 1, 2, \ldots$.

16.15. Write $\Omega_j = \Omega_{p_j-1} = X_j^{p_j-1}$ and let P_j denote the uniform distribution on Ω_j. Let $\Omega = \prod_{j=1}^{\infty} \Omega_j$ be equipped with the probability measure P which is the product of the probability measures P_j on Ω_j. Let $\omega = (\omega_j)_{j=1}^{\infty}$ denote a point of Ω, where $\omega_j \in \Omega_j$; $\omega_j = (x_{j,1}(\omega), \cdots, x_{j,p_j-1}(\omega))$, $x_{j,i}$ being the projection in Ω on the ith component of $X_j^{p_j-1}$.

Let $B_j = \{\omega \in \Omega : \omega_j \notin L_j\}$. Then $P(B_j) \le \frac{1}{10^{j+3}}$, whence $P(B) = 0$, where

$$B = \limsup_{j \to \infty} B_j = \cap_{k=1}^{\infty} \cup_{j=k}^{\infty} B_j.$$

Let $G = \Omega - B$, and call the elements of G the good elements of Ω.

16.16. The following observations are crucial to what will follow:
If $\omega \in G$, then for all except finitely many k, $\omega_k \in L_k$, which in turn implies that there exists k_0 such that for all $k > k_0$,

(i) $| \sum_{i=j}^{j+n} (x_{k,i+1} - x_{k,i})(\omega) | = | x_{k,j+n}(\omega) - x_{k,j}(\omega) | \le h_{k-1}, 1 \le j \le j+n \le p_k - 1$,

(ii) for $n \le (1 - \epsilon_k)(p_k - 1)$ and for any integer l

$$\text{card}\{i : x_{k,i+n}(\omega) - x_{k,i}(\omega) = l\} \le \frac{5}{4}\frac{1}{h_{k-1}}\epsilon_k p_k.$$

(iii) $4\nu(_{k+1}J_1) \le \nu(_kJ_1)\nu(_{k-1}J_1)$.

We would like to show that for a. e. $\omega \in \Omega$, T_ω is mixing. We will show that for every $\omega \in G$, T_ω satisfies,

$$\limsup_{M \to \infty} \nu(T_\omega^M(A) \cap B) \le 20\nu(A)\nu(B)$$

for all $A, B \in \mathcal{B}$. Since for a. e. ω, T_ω is totally ergodic theorem **16.2.** will yield the result.

So fix an $\omega \in G$ and let T denote T_ω. Let us recall the notations. For a stack of height h_k for this T, we denote the levels of this stack by $_kJ_\alpha$, $0 \le \alpha \le h_k - 1$. When we divide this stack into p_k equal columns, the part of

$_k J_\alpha$ which belongs to the ith column is denoted by $_k J_\alpha^i$, $1 \leq i \leq p_k$. We will write T_k for the restriction of T to $\cup_{\alpha=0}^{h_k-2} {_k J_\alpha} = X_k - {_k J_{h_k-1}}$. Note that X_k and T_k are the set and the partial automorphism obtained after the kth stage of the construction of T. The fact that $T = T_\omega$ and that $\omega \in G$ implies that there exists k_0 such that for $k > k_0$ (i)–(iii) above are satisfied. In addition we can assume that for $k > k_0$, $2\nu(_{k-1}J_\beta) \cdot h_{k-1} > 1$, since $\nu(_n J_1) \cdot h_n \to 1$ as $n \to \infty$. We will assume in what follows that $k > k_0$.

16.17 Lemma. *Let the notation be as above. If*
(i) $h_{k-1} < \alpha < h_k - h_{k-1}$
(ii) $0 \leq r < h_{k-1} + h_k$,
(iii) $h_k + 2h_{k-1} < \alpha + r$
then
$$\nu(T^r(_k J_\alpha) \cap {_{k-1}J_\beta}) \leq 9\nu(_k J_\alpha) \cdot \nu(_{k-1}J_\beta),$$
for any level $_{k-1}J_\beta$ of the $(k-1)^{th}$ stack.

Proof. From conditions (i), (ii), (iii) of the lemma we see that,
$$T^r(_k J_\alpha^i) = {_k J_{\alpha+r-h_k-h_{k-1}-(x_{k,i}-x_{k,i-1})(\omega)}^{i+1}}$$

for $0 \leq i \leq p_k - 1$. Observe that, since $\mid (x_{k,i} - x_{k,i-1}) \mid \leq h_{k-1}$, the set $T^r(_k J_\alpha^{i+1})$ lies between the $2h_{k-1} + 1$ consecutive levels
$$_k J_u, \quad \alpha + r - h_k - 2h_{k-1} \leq u \leq \alpha + r - h_k.$$

Let us return to the stack of height h_{k-1}. Fix a level $_{k-1}J_\beta$ of this stack. Note the crucial fact, obvious from construction, that any set of $2h_{k-1} + 1$ consecutive levels of the kth stack will contain at most three levels which are subsets of $_{k-1}J_\beta$. Let $_k J_{\lambda_1}, {_k J_{\lambda_2}}, {_k J_{\lambda_3}}$ be the three levels, if such exist, which are parts of $_{k-1}J_\beta$ and which fall in the above $2h_{k-1} + 1$ consecutive levels
$$_k J_u, \quad \alpha + r - h_k - 2h_{k-1} \leq u \leq \alpha + r - h_k.$$

The number ξ of indices i such that
$$T^r(_k J_\alpha^i) \subseteq {_k J_{\lambda_1}} \cup {_k J_{\lambda_2}} \cup {_k J_{\lambda_3}}$$

is the sum of the cardinalities of the following three sets:
$$\{i : 1 \leq i \leq p_k - 1, \ x_{k,i}(\omega) - x_{k,i-1}(\omega) = \alpha + r - h_k - h_{k-1} - \lambda_1\},$$
$$\{i : 1 \leq i \leq p_k - 1, \ x_{k,i}(\omega) - x_{k,i-1}(\omega) = \alpha + r - h_k - h_{k-1} - \lambda_2\}$$
$$\{i : 1 \leq i \leq p_k - 1, \ x_{k,i}(\omega) - x_{k,i-1}(\omega) = \alpha + r - h_k - h_{k-1} - \lambda_3\}.$$

Since $i - (i-1) = 1 < (1 - \epsilon_k)(p_k - 1)$, (recall that $m_0 = p_k - 1$) and since ω is good, we have
$$\xi < 3 \cdot \frac{5}{4} \cdot \frac{1}{h_{k-1}} \cdot (p_k - 1).$$

Let
$$H = \cup_{1 \le i \le p_k - 1}(_k J_\alpha^i) = {}_k J_\alpha - {}_k J_\alpha^{p_k}.$$

Then
$$\frac{\nu(T^r(H) \cap {}_{k-1}J_\beta)}{\nu(H)} \le \frac{\xi\nu(_k J_\alpha^1)}{(p_k - 1)\nu(_k J_\alpha^1)} \le \frac{15}{2} \cdot \frac{1}{2h_{k-1}} < 8 \cdot \nu(_{k-1}J_\beta)$$

since $2h_{k-1}\nu(_{k-1}J_\beta) > 1$. Thus
$$\nu(T^r(H) \cap {}_{k-1}J_\beta) < 8 \cdot \nu(H) \cdot \nu(_{k-1}J_\beta),$$

and since, by the choice of p_k,
$$\nu(_k J_\alpha^{p_k}) \le \nu(_k J_\alpha) \cdot \nu(_{k-1}J_\beta),$$

we see that,
$$\nu(T^r(_k J_\alpha) \cap {}_{k-1}J_\beta) < 9\nu(_k J_\alpha) \cdot \nu(_{k-1}J_\beta).$$

The lemma is proved.

Remark. Note that $T^r \mid_H = T^r_{k+1} \mid_H$, but whereas T^r is defined on $_k J_\alpha^{p_k}$, T^r_{k+1} is not.

16.18 Lemma. *Fix k and let an integer M satisfy*
$$h_k + h_{k-1} \le M < h_{k+1} + h_k.$$

By division algorithm write
$$M = n(h_k + h_{k-1}) + r, \ n \ge 1, \ 0 \le r < h_k + h_{k-1}$$

Let $_k J_\alpha$ be a level of the kth stack such that

(i) $h_{k-1} < \alpha < h_k - h_{k-1}$,
(ii) $\alpha + r < h_k - h_{k-1}$ or $\alpha + r > h_k + 2h_{k-1}$
(iii) $n > (1 - \epsilon_k)(p_k - 1)$.

Then
$$\nu(T^M(_k J_\alpha) \cap {}_{k-1}J_\beta) \le 12\nu(_k J_\alpha) \cdot \nu(_{k-1}J_\beta),$$

for any level $_{k-1}J_\beta$ of the $(k-1)^{th}$ stack.

Proof. Fix a level $_k J_\alpha$ satisfying
$$h_{k-1} < \alpha < h_k - h_{k-1}.$$

Assume in addition that $\alpha + r < h_k - h_{k-1}$. These conditions will be assumed for α in what follows. (The case when $\alpha + r > h_k + 2h_{k-1}$ is treated similarly.) We have:
$$T^r(_k J_\alpha^i) = {}_k J_{\alpha+r}^i, \ 1 \le i \le p_k,$$

$$T^{h_k+h_{k-1}}\left({}_kJ^i_{\alpha+r}\right) = {}_kJ^{i+1}_{\alpha+r-(x_{k,i}(\omega)-x_{k,i-1}(\omega))}, \quad 1 \le i \le p_k - 1.$$

$$\vdots$$

$$T^{n(h_k+h_{k-1})}{}_kJ^i_{\alpha+r} = {}_kJ^{i+n}_{\alpha+r-(x_{k,i+n-1}(\omega)-x_{k,i-1}(\omega))}, \quad 1 \le i \le p_k - n.$$

Thus

$$T^{n(h_k+h_{k-1})+r}{}_kJ^i_{\alpha} = {}_kJ^{i+n}_{\alpha+r-(x_{k,i+n-1}(\omega)-x_{k,i-1}(\omega))}, \quad 1 \le i \le p_k - n$$

$$T^M{}_kJ^i_{\alpha} = {}_kJ^{i+n}_{\alpha+r-(x_{k,i+n-1}(\omega)-x_{k,i-1}(\omega))}, \quad 1 \le i \le p_k - n$$

Let H be the union of intervals ${}_kJ^i_{\alpha}$, $1 \le i \le \epsilon_k p_k + 1$. Then by the choice of ϵ_k,

$$\nu(H) \le \epsilon_k p_k \nu({}_{k+1}J_1) + \nu({}_{k+1}J_1) \le 2 \cdot \nu({}_kJ_{\alpha}) \cdot \nu({}_{k-1}J_{\beta}).$$

Since $n + \epsilon_k p_k + 1 > p_k + \epsilon_k > p_k$, we see that T^M_{k+1} is not defined on $S = {}_kJ_{\alpha} - H$, which is the union of intervals ${}_kJ^i_{\alpha}, \epsilon_k p_k + 1 < i \le p_k$. These intervals appear as some of the levels in the stack of height h_{k+1}. If ${}_{k+1}J_{\gamma}$ denotes the occurrence of a ${}_kJ^i_{\alpha}$, $\epsilon p_k + 1 < i \le p_k$ in the stack of height h_{k+1}, then $\gamma = (i-1)(h_k + h_{k-1}) + \alpha + x_{k,i-1}$. Further

(a) the fact that T^M_{k+1} is not defined on ${}_{k+1}J_{\gamma}$ means that $\gamma + M \ge h_{k+1}$ hence also $> h_{k+1} - h_k$,

(b) at most three such ${}_{k+1}J_{\gamma}$ appear with γ satisfying

$$h_{k+1} \le \gamma + M \le h_{k+1} + 2h_k$$

since the length of the interval $[h_{k+1}, h_{k+1} + 2h_k]$ is $2h_k + 1$,

(c) at most one such ${}_{k+1}J_{\gamma}$ occurs with $\gamma > h_{k+1} - h_k$,

(d) since $n \ge 1$ and $\alpha > h_{k-1}$, the occurrence of ${}_{k+1}J_{\gamma}$ as ${}_kJ^i_{\alpha}$ implies that $\gamma \ge h_k + h_{k-1} > h_k$.

Let R be the union of (at most four) levels ${}_{k+1}J_{\gamma}$, if such exist, which satisfy (b) or (c). Then by the choice of ϵ_k and p_k,

$$\nu(R) \le \nu({}_kJ_{\alpha}) \cdot \nu({}_{k-1}J_{\beta}).$$

Now let ${}_{k+1}J_{\gamma}$ be a level (which occurs as ${}_kJ^i_{\alpha}$, $\epsilon_k p_k + 1 < i \le p_k$), which is not a subset of R. Then γ satisfies

(i) $h_k < \gamma < h_{k+1} - h_k$,

(ii) $h_{k+1} + 2h_k < \gamma + M$. Moreover,

(iii) $M < h_{k+1} + h_k$.

We can apply the previous lemma, (with k replaced by $k + 1$), and obtain

$$\nu(T^M({}_{k+1}J_{\gamma}) \cap {}_kJ_{\delta}) \le 9\nu({}_{k+1}J_{\gamma}) \cdot \nu({}_kJ_{\delta}),$$

for any level ${}_kJ_{\delta}$ of the kth stack.

Since $_{k-1}J_\beta$ is a union of the levels of the kth stack, we see that

$$\nu(T^M(S) \cap {}_{k-1}J_\beta) \le 9\nu(S) \cdot \nu({}_{k-1}J_\beta),$$

where S is the union of $_{k+1}J_\gamma$ which are not subsets of R (and which occur as $_kJ_\alpha^i$, $\epsilon_k p_k + 1 < i \le p_k$). Finally it is easy to see that

$$\nu(T^M({}_kJ_\alpha) \cap {}_{k-1}J_\beta) = \nu(T^M(H) \cap {}_{k-1}J_\beta) + \nu(T^M(S) \cap {}_{k-1}J_\beta)$$

$$+\nu(T^M(R) \cap {}_{k-1}J_\beta) \le 12\nu({}_kJ_\alpha) \cdot \nu({}_{k-1}J_\beta).$$

This proves the lemma for the case when $\alpha+r < h_k-h_{k-1}$. If $\alpha+r > h_k+2h_{k-1}$ (in addition to $h_{k-1} < \alpha < h_k - h_{k-1}$), then a calculation shows that

$$T^r({}_kJ_\alpha^i) = {}_kJ_{\alpha+r-h_k-h_{k-1}-(x_{k,i}(\omega)-x_{k,i-1}(\omega))}^{i+1}, \quad 1 \le i \le p_k - 1,$$

$$T^{h_k+h_{k-1}}(T^r{}_kJ_\alpha^i) = {}_kJ_{\alpha+r-h_k-h_{k-1}-(x_{k,i+1}(\omega)-x_{k,i-1}(\omega))}^{i+2}, \quad 1 \le i \le p_k - 2$$

$$\vdots$$

$$T^{n(h_k+h_{k-1})+r}({}_kJ_\alpha^i) = {}_kJ_{\alpha+r-h_k-h_{k-1}-(x_{k,i+n}(\omega)-x_{k,i-1}(\omega))}^{i+n+1},$$

$1 \le i \le p_k - n - 1$. Thus

$$T^M({}_kJ_\alpha^i) = {}_kJ_{\alpha+r-h_k-h_{k-1}-(x_{k,i+n}(\omega)-x_{k,i-1}(\omega))}^{i+n+1},$$

$1 \le i \le p_k - n - 1$.

Since $n > (1-\epsilon_k)(p_k - 1)$, $(n+1) > (1-\epsilon_k)(p_k - 1)$, and the calculations of the previous case can be repeated. The lemma is proved.

16.19 Lemma. *Fix a k and let M be an integer satisfying*

$$h_k + h_{k-1} \le M < h_{k+1} + h_k.$$

Then we can delete a certain number of levels from the stack of height h_k such that

(i) *the total measure of the deleted intervals is less than 10^{-k+3},*
(ii) *if $_kJ_\alpha$ is any level not in the deleted levels, and if $_{k-1}J_\beta$ is any level of the stack of height h_{k-1}, then*

$$\nu(T^M({}_kJ_\alpha) \cap {}_{k-1}J_\beta) \le 17\nu({}_kJ_\alpha) \cdot \nu({}_{k-1}J_\beta).$$

Proof. By division algorithm we write

$$M = n(h_k + h_{k-1}) + r, \quad n \ge 1, \quad 0 \le r < h_k + k_{k-1}.$$

Delete from the stack of height h_k the top $h_{k-1}+1$ levels, the bottom $h_{k-1}+1$ levels and the levels $_kJ_\alpha$ with α satisfying

$$\alpha + r \in [h_k - h_{k-1}, h_k + 2h_{k-1}].$$

The number of deleted levels is less than $6h_{k-1}$ and so the total measure of these levels is less than

$$6\nu(_kJ_1)h_{k-1} \leq \frac{6}{h_k} \cdot h_{k-1} \leq \frac{6}{10^k} < 10^{-k+3}.$$

Let $_kJ_\alpha$ be a level not among the deleted levels. Then α satisfies:

(i) $h_{k-1} < \alpha < h_{k-1} - h_{k-1}$,
(ii) $\alpha + r < h_k - h_{k-1}$ or $\alpha + r > h_k + 2h_{k-1}$.

In case $n > (1 - \epsilon_k)(p_k - 1)$, the lemma follows from the previous lemma. So we assume that $n \leq (1 - \epsilon_k)(p_k - 1)$ and consider first the case when $\alpha + r < h_k - h_{k-1}$, (in addition to $h_{k-1} < \alpha < h_k - h_{k-1}$). As before we have

$$T^M(_kJ_\alpha^i) = {}_kJ_{\alpha+r-(x_{k,i+n-1}(\omega)-x_{k,i-1}(\omega))}^{i+n}, \quad 1 \leq i \leq p_k - n$$

and $T^M(_kJ_\alpha^i)$ is a subset of the $2h_{k-1} + 1$ consecutive levels

$$_kJ_u, \quad \alpha + r - h_{k-1} \leq u \leq \alpha + r + h_{k-1}.$$

Let us return to the stack of height h_{k-1} and fix a level $_{k-1}J_\beta$ of this stack. Let $_kJ_{\lambda_1}, {}_kJ_{\lambda_2}, {}_kJ_{\lambda_3}$ be the three levels, if such exist, which are parts of $_{k-1}J_\beta$ and which fall in the above $2h_{k-1} + 1$ consecutive levels

$$_kJ_u, \quad \alpha + r - h_{k-1} \leq u \leq \alpha + r + h_{k-1}.$$

The number ξ of indices i such that

$$T^M(_kJ_\alpha^i) \subseteq {}_kJ_{\lambda_1} \cup {}_kJ_{\lambda_2} \cup {}_kJ_{\lambda_3}$$

is the sum of the cardinalities of the following three sets:

$$\{i : 1 \leq i \leq p_k - n, \; x_{k,i+n-1}(\omega) - x_{k,i-1}(\omega) = \alpha + r - \lambda_1\}$$

$$\{i : 1 \leq i \leq p_k - n, \; x_{k,i+n-1}(\omega) - x_{k,i-1}(\omega) = \alpha + r - \lambda_2\}$$

$$\{i : 1 \leq i \leq p_k - n, \; x_{k,i+n-1}(\omega) - x_{k,i-1}(\omega) = \alpha + r - \lambda_3\}$$

Since $i + n - 1 - (i - 1) = n \leq (1 - \epsilon_k)(p_k - 1)$ and since ω is good, we have

$$\xi < 3 \cdot \frac{5}{4} \cdot \frac{1}{h_{k-1}} \cdot (p_k - n - 1) < \frac{15}{4} \cdot \frac{1}{h_{k-1}}(p_k - n).$$

(Recall that $m_0 = p_k - 1$.) Let

$$H = \cup_{1 \leq i \leq p_k - n}(_kJ_\alpha^i) = {}_kJ_\alpha - \cup_{p_k - n < i \leq p_k}(_kJ_\alpha^i).$$

Then

$$\frac{\nu(T^M(H) \cap {}_{k-1}J_\beta)}{\nu(H)} \leq \frac{\xi\nu(_kJ_\alpha^1)}{(p_k - n)\nu(_kJ_\alpha^1)}$$

$$\leq \frac{15}{4} \cdot \frac{1}{h_{k-1}} = \frac{15}{2} \cdot \frac{1}{2h_{k-1}}$$

$$< 8 \cdot \nu(_{k-1}J_\beta),$$

since $2h_{k-1}\nu(_{k-1}J_\beta) > 1$. Thus

$$\nu(T^r(H) \cap {}_{k-1}J_\beta) < 8 \cdot \nu(H) \cdot \nu(_{k-1}J_\beta). \tag{1}$$

Consider now the action of T^M on the set $S = {}_kJ_\alpha - H$, the union of intervals $_kJ_\alpha^i$, $p_k - n < i \leq p_k$. Since now $i + n > p_k$, we see that T_{k+1} is not defined on S and we can repeat the argument of the last lemma (going over to the stack of height $h_{k+1}etc.$), and conclude that

$$\nu(T^M(S) \cap {}_{k-1}J_\beta) \leq 9\nu(_kJ_\alpha)\nu(_{k-1}J_\beta).$$

Combining this with (1) above we see that if $\alpha + r < h_k - h_{k-1}$, then

$$\nu(T^M(_kJ_\alpha) \cap {}_{k-1}J_\beta) \leq 17\nu(_kJ_\alpha)\nu(_{k-1}J_\beta).$$

The case when $\alpha + r > h_k + 2h_{k-1}$ is treated similarly except that we have to consider the cases $n + 1 \leq (1 - \epsilon_k)(p_k - 1)$ and $n + 1 > (1 - \epsilon_k)(p_k - 1)$. The lemma is proved.

16.20. Let $\mathcal{B}_k$ denote the collection of sets which are unions of levels of the stack of height h_k. Then $\mathcal{B}_k \subset \mathcal{B}_{k+1}$. Let $\mathcal{C} = \cup_{k=1}^\infty \mathcal{B}_k$ and let $A, B \in \mathcal{C}$. Then $A, B \in \mathcal{B}_k$ for some k. We may then assume that $A \in \mathcal{B}_{k+n}, B \in \mathcal{B}_{k+n-1}$ for all integers $n \geq 1$. It is easy to see that the last lemma implies

$$\limsup_{M \to \infty} \nu(T^M(A) \cap B) \leq 17\nu(A) \cdot \nu(B).$$

Since $\mathcal{C}$ generates the σ-algebra $\mathcal{B}$ (unto ν-null sets) we have proved

16.21 Theorem. *If $\omega \in G$ then for all $A, B \in \mathcal{B}$,*

$$\limsup_{M \to \infty} \nu(T^M(A) \cap B) \leq 17\nu(A) \cdot \nu(B).$$

Chapter 17

Calculus of Generalized Riesz Products

17.1. In this chapter we discuss generalized Riesz products bringing into consideration the H^p theory, the notion of Mahler measure, and the zeros of polynomials appearing in the generalized Riesz product. A natural Formula for Radon-Nikodym derivative between two generalized Riesz products is established under suitable conditions. This is then used to formulate a dichotomy theorem and prove a conditional version of it. A discussion involving spectrum of non-singular rank one maps and flat polynomials is given. (El. Abdalaoui, Nadkarni [4])

Generalized Riesz Products and their Weak Dichotomy

In this section we introduce generalized Riesz products and derive a weak dichotomy result about the infinite sequence of polynomials associated to it. This also yields conditions for absolute continuity of generalized Riesz products.

17.2 Definition. Let $P_1, P_2, \cdots$, be a sequence of trigonometric polynomials such that

1. for any finite sequence $i_1 < i_2 < \cdots < i_k$ of natural numbers

$$\int_{S^1} \left| (P_{i_1} P_{i_2} \cdots P_{i_k})(z) \right|^2 dz = 1,$$

   where S^1 denotes the circle group and dz the normalized Lebesgue measure on S^1,

2. for any infinite sequence $i_1 < i_2 < \cdots$ of natural numbers the weak limit of the measures $| (P_{i_1} P_{i_2} \cdots P_{i_k})(z) |^2 dz, k = 1, 2, \cdots$ as $k \to \infty$ exists. Then the measure μ given by the weak limit of $| (P_1 P_2 \cdots P_k)(z) |^2 dz, k =$

$1, 2, \ldots$ as $k \to \infty$ is called generalized Riesz product of the polynomials $\mid P_1 \mid^2, \mid P_2 \mid^2, \cdots$ and denoted by

$$\mu = \prod_{j=1}^{\infty} |P_j|^2 \qquad (1).$$

For an increasing sequence $k_1 < k_2 < \cdots$ of natural numbers the product $\prod_{j=1}^{\infty} |P_{k_j}|^2$ makes sense as the weak limit of probability measures $|(P_{k_1} P_{k_2} \cdots P_{k_n})(z)|^2 dz$ as $n \to \infty$. It depends on the sequence $k_1 < k_2 \cdots$, and called a subproduct of the given generalized Riesz product.

Since the object under consideration is the generalized Riesz product $\prod_{j=1}^{\infty} |P_j|^2$, without loss of generality we assume that the polynomials $P_j, j = 1, 2, \cdots$ are analytic with positive constant term. Their domain of definition will mainly be the circle group, but with option to look at then as functions on the complex plane. Since $\int_{S^1} |P_j|^2(z) dz = 1$, the sum of the squares of the absolute values of coefficients of P_j is one, and so each coefficient of P_j is at most one in absolute value. Let $a_0^{(j)}$ denote the constant term of P_j, which is positive by assumption. The sequence of products $\prod_{j=1}^{n} a_0^{(j)}, n = 1, 2, \cdots$ is non-increasing, and so has a limit which is either zero or some positive constant which can be at most 1. The case when this constant is one is obviously the trivial case when each P_j is the constant 1.

Consider the sequence of polynomials $S_n \stackrel{\text{def}}{=} \prod_{j=1}^{n} P_j, n = 1, 2, \cdots$ (without the absolute value squared). For each n, let $b_0^{(n)} \stackrel{\text{def}}{=} \prod_{j=1}^{n} a_0^{(j)}$ denote the constant term of S_n. Write $b = \lim_{n \to \infty} b_0^{(n)}$. We have the following weak dichotomy theorem for generalized Riesz products.

17.3 Theorem. *If $b = 0$, the sequence of polynomials $S_n = \prod_{i=1}^{n} P_i, i = 1, 2, \cdots$ converges to zero weakly in $L^2(S^1, dz)$. If b is positive it converges in $L^1(S^1, dz)$ (and in H^1) to a non-zero function f which is also in H^2 with H^2 norm at most 1, $\log(|f|)$ has finite integral.*

Proof Assume that $b = 0$. We show that the sequence $S_n, n = 1, 2, \cdots$ has zero weak limit as functions in $L^2(S^1, dz)$. Assume that a subsequence $S_{k_n}, n = 1, 2, \cdots$ converges weakly to a function $f \in L^2(S^1, dz)$. We show that f is the zero function. By choosing a further subsequence if necessary we can assume without any loss that the constant term of $\frac{S_{k_{n+1}}}{S_{k_n}}, n = 1, 2, \cdots$ goes to zero as $n \to \infty$. Since $b = 0$, the zeroth Fourier coefficient of f is zero. Since each S_n is an analytic trigonometric polynomial, the negative Fourier coefficients of f are all zero. Assume now that for $0 \leq j < l$, $b_j = \int_{S^1} z^{-j} f(z) dz = 0$. Then, given $\epsilon > 0$, for large enough m, $\left| \int_{S^1} z^{-j} S_{k_m} dz \right| < \epsilon$, for $0 \leq j < l$, and moreover the constant term of $\frac{S_{k_{m+1}}}{S_{k_m}}$ is less than ϵ. For $n > m$,

$$\prod_{j=1}^{k_n} P_i = \prod_{j=1}^{k_m} P_j \prod_{j=k_m+1}^{k_n} P_j.$$

Since P_j's are one sided trigonometric polynomials, it is easy to see from this that $\left| \int_{S^1} z^{-l} S_{k_n}(z)dz \right| \leq (l+1)\epsilon$. Since this holds for all $n > m$ we see that $\int_{S^1} z^{-l} f(z)dz = 0$. Induction completes the proof.

Assume now that $b > 0$. Then $a_0^{(n)}$ as well as $\prod_{j=m+1}^n a_0^{(j)}, m < n$, converge to 1 as $m, n \to \infty$. Since $L^2(S^1, dz)$ norm of all the finite products is one,

$$P_n, \frac{S_n}{S_m} \to 1 \quad \text{in} \quad L^2(S^1, dz), \quad m < n, \quad \text{as} \quad m, n \to \infty.$$

Moreover by Cauchy-Schwarz inequality

$$\left\| S_n - S_m \right\|_1 = \left\| S_m \left(1 - \prod_{m+1}^n P_i \right) \right\|_1 \leq \left\| S_m \right\|_2 \left\| 1 - \prod_{j=m+1}^n P_j \right\|_2$$

$$\to 0 \quad \text{as} \quad m, n \to \infty.$$

Thus the sequence of analytic polynomials $S_n, n = 1, 2, \cdots$ converges in $L^1(S^1, dz)$ to a function which we denote by f, and view it also as a function in the Hardy class H^1. A subsequence of $S_n, n = 1, 2, \cdots$ converges to f a.e (with respect to the Lebesgue measure of S^1), whence, over the same subsequence $S_n^2, |S_n|^2, n = 1, 2, \cdots$ converge to f^2 and $|f|^2$, respectively. Since $\|S_n\|_2 = 1, n = 1, 2, \cdots$, by Fatou's lemma we conclude that f is square integrable with $L^2(S^1, dz)$ norm at most 1. Thus f is in H^2, and $\log |f|$ has finite integral. This completes the proof of the theorem.

We do not know if the $L^2(S^1, dz)$ norm of f is 1, equivalently, if $S_n, n = 1, 2, \cdots$ converges to f in $L^2(S^1, dz)$. We give some sufficient conditions under which this holds. Let $S_n = \sum_{j=0}^{m_n} b_j^{(n)} z^j$, where m_n is the degree of the trigonometric polynomial S_n. Now

$$b_j^{(n)} = \int_{S^1} z^{-j} S_n(z)dz \to \int_{S^1} z^{-j} f(z)dz \overset{\text{def}}{=} b_j.$$

The series $\sum_{j=0}^\infty b_j z^j$ is the Fourier series of f and we call this series the formal expansion of $\prod_{j=1}^\infty P_j$. Since b is positive, the infinite product $\prod_{j=n+1}^\infty a_0^{(j)}$ is also positive, so the infinite product $\prod_{j=n+1}^\infty P_j$ has a formal expansion which we denote by $\sum_{j=0}^\infty c_j^{(n)} z^j$. Note that $c_0^{(n)} = \prod_{j=n+1}^\infty a_0^{(j)} \longrightarrow 1$ as $n \longrightarrow \infty$, as a result $\sum_{j=1}^\infty |c_j^{(n)}|^2 \leq 1 - (c_0^{(n)})^2 \longrightarrow 0$ as $n \to \infty$.

At this point we recall the definition of dissociated polynomials (see **15.3-15.6**).

17.4 Definition. Finitely many trigonometric polynomials $q_0, q_1, \cdots, q_n$, $q_j = \sum_{i=-N_j}^{N_j} d_i^{(j)} z^i$, $j = 0, 1, 2, \cdots, n$ are said to be dissociated if in their product $q_0(z) q_1(z) \cdots q_n(z)$, (when expanded formally, i.e., without grouping terms or canceling identical terms with opposite signs), the powers $i_0 + i_1 + \cdots + i_n$ of z in non-zero terms

$$d_{i_0}^{(0)} d_{i_1}^{(1)} \cdots d_{i_n}^{(n)} z^{i_0 + i_1 + \cdots + i_n}$$

are all distinct. A sequence $q_0, q_1, \cdots$, of trigonometric polynomials is said to be dissociated if for each n the polynomials $q_0, q_1, \cdots, q_n$ are dissociated.

Suppose now that the polynomials $P_1, P_2, \cdots$ (without the squares) appearing in generalized Riesz product (1) are dissociated. Then, whenever $b_j^{(n)}$ is a non-zero coefficient in the expansion of S_n, $b_j^{(l)} = b_j^{(n)} \frac{b_0^{(l)}}{b_0^{(n)}}$ for all $l \geq n$. Thus, if the polynomials P_j, $j = 1, 2, \cdots$ are dissociated, then we see on letting $l \to \infty$ that $b_j = b_j^{(n)} c_0^{(n)}$, provided $b_j^{(n)} \neq 0$. We therefore have for any n

$$\sum_{j=0}^{m_n} |b_j|^2 \geq \sum_{j=1}^{m_n} |b_j^{(n)}|^2 (c_0^{(n)})^2 = (c_0^{(n)})^2 \longrightarrow 1$$

as $n \to \infty$. Thus f has $L^2(S_1, dz)$ norm 1. We have proved:

17.5 Theorem. *If the polynomials $P_n, n = 1, 2, \cdots$ are dissociated and b is positive then the partial products $S_n, n = 1, 2, \cdots$ converge in H^2 to a non-zero function f and the generalized product $\prod_{j=1}^{\infty} |P_j|^2$ is the measure $|f|^2 dz$. Further, $\int_{S^1} \log(|f(z)|) dz$ is finite.*

If we replace the condition that the polynomials $P_n, n = 1, 2, \cdots$ are dissociated by the condition that coefficients of the polynomials $P_n, n = 1, 2, \cdots$ are all non-negative, then it is easy to verify that for $0 \leq k \leq m_n$,

$$b_k \geq c_0^{(n)} b_k^{(n)} + b_0^{(n)} c_k^{(n)},$$

whence

$$\sum_{k=0}^{\infty} |b_k|^2 \geq \left(\sum_{k=0}^{m_n} |b_k^{(n)}|^2 \right) |c_0^{(n)}|^2 = 1 \cdot |c_0^{(n)}|^2 \to 1,$$

as $n \to \infty$. Thus, if the coefficients of all the $P_n, n = 1, 2, \cdots$ are non-negative, and if $b = b_0 > 0$, we necessarily have convergence of $S_n, n = 1, 2, \cdots$ in H^2.

We continue with the assumption that b is positive, but no more assume that the polynomials $P_n, n = 1, 2, \cdots$ are dissociated or have non-negative coefficients. Fix n, and let $1 \leq j \leq m_n$, then

$$\sum_{j=0}^{\infty} b_j z^j = \left(\sum_{i=0}^{m_n} b_i^{(n)} z^i \right) \left(\sum_{k=0}^{\infty} c_k^{(n)} z^k \right).$$

This gives for any $j \geq 0$,

$$b_j = b_j^{(n)} c_0^{(n)} + \sum_{i=0}^{j-1} b_i^{(n)} c_{j-i}^{(n)}.$$

Hence, for any $j \geq 1$,

$$| b_j - b_j^{(n)} c_0^{(n)} |^2 \leq \left(\sum_{i=0}^{j-1} | b_i^{(n)} |^2 \right) \left(\sum_{i=0}^{j-1} | c_{j-i}^{(n)} |^2 \right) \leq \sum_{j=1}^{\infty} | c_j^{(n)} |^2$$

$$\leq 1 - (c_0^{(n)})^2 \longrightarrow 0 \quad \text{as} \quad n \to \infty$$

Assume now that $m_n (1 - c_0^{(n)}) \longrightarrow 0$ as $n \to \infty$. Then $\sum_{j=0}^{m_n} | b_j - b_j^{(n)} c_0^{(n)} |^2 \to 0$ as $n \to \infty$. Another use of the assumption that $m_n (1 - c_0^{(n)}) \longrightarrow 0$ as $n \to 0$ allows us to conclude that $\sum_{j=0}^{m_n} |b_j - b_j^{(n)}|^2 \to 0$ as $n \to \infty$. Since $\sum_{j=1}^{m_n} |b_j^{(n)}|^2 = 1$, we conclude that $\sum_{j=1}^{\infty} |b_j|^2 = 1$, so that $L^2(S^1, dz)$ norm of f is one and $S_n, n = 1, 2, \cdots$ converges to f in H^2. We have proved:

17.6 Theorem. *If b is positive and $m_n(1 - c_0^{(n)}) \longrightarrow 0$ as $n \to \infty$, then $S_n \to f$ in H^2 and $|f|^2 dz$ is the generalized Riesz product $\prod_{j=1}^{\infty} |P_j|^2$. Moreover $\log(|f|)$ has finite integral.*

Exercise. Prove that the above theorem is valid if we require only that $l_n(1 - c_0^n) \to 0$ as $n \to 0$ where, for each n, l_n is the number of non-vanishing terms in S_n.

17.7 Corollary. *If $b > 0$ then there is always a subproduct $\prod_{k=1}^{\infty} P_{n_k}$ for which the condition of the above theorem is satisfied, so that if $b > 0$ holds, then a subproduct $\prod_{k=1}^{\infty} | P_{n_k}(z) |^2$ has the same null sets as Lebesgue measure.*

Proof. Put $k_1 = 1$ and $P_{k_1} = P_1$. Let m_1 be the degree of P_{k_1}. Since $b > 0$, $c_0^{(n)} \to 1$ as $n \to \infty$. Choose $k_2 > k_1$ such that $m_1(1 - c_0^{(k_2)}) \leq \frac{1}{2}$. Consider $P_{k_1} \cdot P_{k_2+1}$. Suppose its degree is m_2. Choose $k_3 > k_2$ such that $m_2(1 - c_0^{(k_3)}) \leq \frac{1}{4}$. Assume that we have chosen $k_1 < k_2 < \cdots < k_{l-1}$ such that for any $i, 1 \leq i \leq l - 2$ if m_i is the degree of $P_{k_1} P_{k_2+1} \cdots P_{k_i+1}$, then

$$m_i \left(1 - c_0^{(k_{(i+1)})} \right) \leq \frac{1}{2^i}.$$

Choose $k_l > k_{l-1}$ such that

$$m_{l-1} \left(1 - c_0^{(k_l)} \right) \leq \frac{1}{2^{l-1}}.$$

Thus we have inductively chosen a sequence $k_1 < k_2 < k_3 < \cdots < k_i < \cdots$. Write $J_1 = P_{k_1}, J_2 = P_{k_2+1}, \cdots, J_n = P_{k_n+1}, \cdots$ and $R = \prod_{i=0}^{\infty} | J_i(z) |^2$. If γ_n

denotes the constant term of $\prod_{i=n+1}^{\infty} J_i$, then it is easy to see that $\gamma_n > c_0^{(k_n)}$ so that $p_n(1 - \gamma_{n+1}) \leq \frac{1}{2^n}$, where p_n is the degree of $\prod_{i=1}^{n} J_i$. By the theorem above the Riesz product $R = \prod_{i=1}^{\infty} | J_i(z) |^2$ is equivalent to the Lebesgue measure. This completes the proof of the corollary.

Assume that b is positive. Consider $L(z) = \prod_{j=m+1}^{n} |P_j(z)|^2$. If $d_k(m,n) = d_k$ is the coefficient of z^k in $\prod_{j=m+1}^{n} P_j$, then for $k > 0$, the coefficient of z^k in $L(z)$ is in absolute value

$$\left| \sum_{j\geq k} d_j \overline{d_{j-k}} \right| \leq \left(\sum_{j\geq k} | d_j |^2 \right)^{\frac{1}{2}} \leq (1 - d_0^2)^{\frac{1}{2}}.$$

Under the assumption that b is positive we can make this coefficient (which depends on m and n) as small as we please by choosing m large. We have proved:

17.8 Theorem. *If b is positive, the generalized Riesz products $\mu_n \stackrel{def}{=} \prod_{n+1}^{\infty} | P_i |^2$, $n = 1, 2, \cdots$ converge weakly to the Lebesgue measure on S^1 as $n \to \infty$.* We do not know if the conclusion of the above theorem holds also when b is zero, but such generalized Riesz products form an important class of measures and will be discussed later (see **17.19**).

17.9 Remark. The weak dichotomy theorem is rather weak in the sense that no information can be garnered about μ, such as absolute continuity or singularity, when b is zero. Consider the classical Riesz product

$$\mu = \prod_{j=1}^{\infty} | \cos(\theta_j) + \sin(\theta_j) z^{n_j} |^2, \quad \frac{n_{j+1}}{n_j} \geq 3, \quad 0 < \theta_j < \frac{\pi}{2}, \quad j = 1, 2, \cdots$$

It is known to be absolutely continuous if $\sum_{j=1}^{\infty} \cos^2(\theta_j) \sin^2(\theta_j)$ is finite and singular otherwise. Clearly the condition for absolute continuity is satisfied with $\cos(\theta_j) = \frac{1}{j}, j = 1, 2, \cdots$ and also with $\cos(\theta_j) = \sqrt{1 - \left(\frac{1}{j}\right)^2}$, $j = 1, 2, \cdots$. In the first case the product of the constant terms is zero, while in the second case it is positive. This defect is rectified if we replace the polynomials P_j with their outer parts, as discussed in the next section.

Outer Polynomials and Mahler Measure

Let

$$\mu = \prod_{j=1}^{\infty} | P_j(z) |^2 \tag{1}$$

be a generalized Riesz product. Let μ_a denote the part of μ absolutely continuous with respect to dz. We write $\frac{d\mu}{dz}$, to mean $\frac{d\mu_a}{dz}$. In this section we use the classical prediction theoretic ideas to evaluate $\exp\left(\int_{S^1}\log\left(\frac{d\mu}{dz}\right)dz\right)$ a quantity which we call the Mahler measure of μ (denoted by $M(\mu)$) with respect to the Lebesgue measure. We will prove:

17.10 Theorem.

$$\int_{S^1}\log\frac{d\mu_a}{dz}dz = \lim_{n\to\infty}\int_{S^1}\log\prod_{j=1}^{n}\mid P_j(z)\mid^2 dz.$$

Note that the theorem is false if we drop the log on both sides of the equation, for then the right hand side is always one, while the left hand side is zero for μ singular to Lebesgue measure. For the proof we begin by recalling Beurling's inner and outer factors for the case of polynomials and the expression for one step 'prediction error', namely the quantity:

$$\inf_{q\in\mathcal{Q}}\int_{S^1}\mid 1-q(z)\mid^2\mid P(z)\mid^2 dz,$$

where $P(z)$ is an analytic trigonometric polynomial with $L^2(S^1,dz)$ norm 1 and non-zero constant term. $\mathcal{Q}$ is the class of all analytic trigonometric polynomials with zero constant term. To this end we have to bring into consideration the zeros of polynomials $P_j, j = 1, 2, \cdots$. Consider the kth polynomial of the generalized Riesz product product $\prod_{k=1}^{\infty}\mid P_k\mid^2$. Suppressing the index k, it is of the type:

$$P(z) = a_0 + a_1 z + \cdots + a_m z^m.$$

assuming that it is of degree m. Let

$$A = \{a : P(a) = 0, \mid a\mid< 1\},$$

$$B = \{b : P(b) = 0, \mid b\mid= 1\},$$

$$C = \{c : P(c) = 0, \mid c\mid> 1\}.$$

Then

$$P(z) = a_m\prod_{a\in A}(z-a)\prod_{b\in B}(z-b)\prod_{c\in C}(z-c)$$

$$= \prod_{a\in A}\frac{(z-a)}{(1-\bar{a}z)}a_m\prod_{a\in A}(1-\bar{a}z)\prod_{b\in B}(z-b)\prod_{c\in C}(z-c).$$

Write

$$I(z) = \bar{\gamma}\prod_{a\in A}\frac{(z-a)}{(1-\bar{a}z)},$$

$$O(z) = \gamma a_m\prod_{a\in A}(1-\bar{a}z)\prod_{b\in B}(z-b)\prod_{c\in C}(z-c).$$

where γ is a constant of absolute value 1 such that the constant term of $O(z)$ is positive, while $\overline{\gamma}$ is the complex conjugate of γ. We have,

$$P(z) = I(z)O(z).$$

Note that for $z \in S^1$, $| I(z) | = 1$, $| P(z) | = | O(z) |$. The function $O(z)$ is non-vanishing inside the unit disc. The functions I and O are Beurling's inner and outer parts of the polynomial P. Note that, since constant term of P is non-zero, the degree of O is same as that of P and that $O(0) = $ constant term of $O \geq P(0) = a_0$. Recall that outer functions in H^2 are precisely those functions f in H^2 for which the functions $z^n f, n \geq 0$, span H^2 in the closed linear sense. Hence, if f is an outer function in H^2, then the closed linear span of $\{z^n f, n \geq 1\}$ is zH^2. The orthogonal projection of $O(z)$ on zH^2 is $O(z) - O(0)$ where $O(0)$ is the constant term of $O(z)$ which we denote by α.

We have

$$|\alpha|^2 = \int_{S^1} |\alpha|^2 \, dz = \int_{S^1} \left| (O(z) - (O(z) - \alpha) \right|^2 dz$$

$$= \int_{S^1} \left| 1 - \frac{(O(z) - \alpha)}{O(z)} \right|^2 |O(z)|^2 \, dz$$

$$= \inf_{q \in \mathcal{Q}} \int_{S^1} |1 - q(z)|^2 |O(z)|^2 \, dz,$$

$$= \inf_{q \in \mathcal{Q}} \int_{S^1} |1 - q(z)|^2 |P(z)|^2 \, dz$$

where the infimum is taken over the class $\mathcal{Q}$ of all analytic trigonometric polynomials q with zero constant term. Thus $\frac{O(z) - \alpha}{O(z)}$ is the orthogonal projection of the constant function 1 on the closed linear span of $\{z^n, n \geq 1\}$ in $L^2(S^1, |P(z)|^2 dz)$.

17.11 Lemma. *If λ is a probability measure on S^1 such that $d\nu = | P(z) |^2 d\lambda$ is again a probability measure then*

$$| \alpha |^2 \geq \inf_{q \in \mathcal{Q}} \int_{S^1} | 1 - q(z) |^2 \, d\nu.$$

Proof. If $O(z)$ has no zeros on the unit circle then $\dfrac{O(z) - \alpha}{O(z)}$ is analytic on the closed unit disk. The partial sums of the power series of this function converge to it uniformly on the unit circle. Let $q_k, k = 1, 2, \cdots$ be the sequence of these partial sums. Then

$$| \alpha |^2 = \int_{S^1} \left| 1 - \frac{O(z) - \alpha}{O(z)} \right|^2 |O(z)|^2 \, d\lambda$$

$$= \int_{S^1} \left| 1 - \frac{O(z) - \alpha}{O(z)} \right|^2 d\nu$$

$$= \lim_{k \to \infty} \int_{S^1} |1 - q_k|^2 \, d\nu \geq \inf_{q \in \mathcal{Q}} \int_{S^1} |1 - q(z)|^2 \, d\nu$$

This conclusion remains valid even if $O(z)$ has zeros on the circle but the proof is slightly different. For fixed r, $0 \leq r < 1$, the function $\frac{O(z) - \alpha}{O(rz)}$ is analytic on the closed unit disk, so the partial sums of its power series converge to it uniformly on S^1. Now for any fixed real θ, $\dfrac{z - e^{i\theta}}{rz - e^{i\theta}}$ remains bounded by 2 for $z \in S^1$ and $0 \leq r < 1$, and converges to 1 as $r \to 1$, for $z \neq e^{i\theta}$. Therefore $z \neq e^{i\theta}$, $\dfrac{O(z)}{O(rz)} \to 1$ boundedly as $r \to 1$, whence

$$\left| 1 - \frac{O(z) - \alpha}{O(rz)} \right|^2 |O(z)|^2 \to |\alpha|^2$$

boundedly as $r \to 1$. It is easy to see from this that

$$\inf_{q \in \mathcal{Q}} \int_{S^1} |1 - q|^2 \, d\nu \leq \lim_{r \to 1} \int_{S^1} \left| 1 - \frac{O(z) - \alpha}{O(rz)} \right|^2 \, d\nu = |\alpha|^2 .$$

This proves the lemma.

Consider now the polynomials P_k, $k = 1, 2, \cdots$ and the associated finite products $\prod_{k=1}^n P_k$, $n = 1, 2, 3, \cdots$. Let $A_k, B_k, C_k, I_k, O_k, \alpha_0^{(k)}$ have the obvious meaning: they are for P_k what A, B, C, I, O, α are for P. Note that the inner and outer parts of $\prod_{k=1}^n P_k$ are $\prod_{k=1}^n I_k$ and $\prod_{k=1}^n O_k$ respectively and the constant term of the outer part is $\prod_{k=1}^n \alpha_0^{(k)}$. Note that $\prod_{k=1}^\infty \alpha_0^{(k)} \overset{def}{=} \beta \geq \prod_{k=1}^\infty a_0^{(k)} = b$, so if b is positive, then so is β. On the other hand, the positivity of β does not in general imply positivity of b as shown by the case of classical Riesz product, see remark (**17.13**(1) below).

To prove Theorem **17.10** we apply the lemma above to

$$P = \prod_{k=1}^n P_k(z)$$

and

$$\lambda = \prod_{k=n+1}^\infty |P_k(z)|^2,$$

Note that

$$d\nu = \left(\prod_{k=1}^n |P_k(z)|^2 \right) d\lambda = d\mu.$$

We see that for any n

$$\inf_{q \in \mathcal{Q}} \int_{S^1} |1 - q|^2 \, d\mu \leq \prod_{k=1}^n |\alpha_0^{(k)}|^2,$$

whence

$$\inf_{q \in \mathcal{Q}} \int_{S^1} |1 - q|^2 \, d\mu \le \prod_{k=1}^{\infty} |\alpha_0^{(k)}|^2 \, .$$

To prove the above inequality in the reverse direction we note that by Szegö's theorem as generalized by Kolmogorov and Krein (see K. H. Hoffman [12]) that

$$\exp\left\{ \int_{S^1} \log\left(\frac{d\mu_a}{dz}\right) dz \right\} = \inf_{q \in \mathcal{Q}} \int_{S^1} |1 - q(z)|^2 \, d\mu.$$

Denote this infimum by l. Then, given $\epsilon > 0$, there is a polynomial $q_0 \in \mathcal{Q}$ such that

$$l \le \int_{S^1} |1 - q_0|^2 \, d\mu < l + \epsilon,$$

whence for large enough n

$$\int_{S^1} |1 - q_0|^2 \prod_{k=1}^{n} |P_k|^2 \, dz < l + \epsilon.$$

Since

$$\left| \prod_{k=1}^{n} \alpha_0^{(k)} \right|^2 = \inf_{q \in \mathcal{Q}} \int_{S^1} |1 - q|^2 \prod_{k=1}^{n} |P_k(z)|^2 \, dz,$$

we see that $|\alpha_0^{(1)} \cdot \alpha_0^{(2)} \cdots \alpha_0^{(n)}|^2 \le l + \epsilon$. Since ϵ is arbitrary positive real number, and $|\alpha_0^{(k)}| < 1$ for all k, we have

$$\prod_{k=1}^{\infty} |\alpha_0^{(k)}|^2 \le l.$$

We also note that

$$\prod_{j=1}^{n} |\alpha_0^{(j)}| = \exp\left\{ \int_{S^1} \log\left(\left| \prod_{j=1}^{n} P_j(z) \right| \right) dz \right\}$$

Thus we have proved:

$$\prod_{k=1}^{\infty} |\alpha_0^{(k)}|^2 = \exp\left\{ \int_{S^1} \log\left(\frac{d\mu_a}{dz}\right) dz \right\}$$

$$= \lim_{n \to \infty} \exp\left\{ \int_{S^1} \log\left(\prod_{j=1}^{n} |P_j(z)|^2 \right) dz \right\}.$$

which is indeed theorem **17.10** with some additional information.

17.12 Corollary. *If each $P_i, i = 1, 2, \cdots$ is outer, then $\log\left(\frac{d\mu}{dz}\right)$ has finite integral if and only if β is positive.*

17.13 Remarks.

1. For the trigonometric polynomial $P_j(z) = \cos\theta_j + \sin\theta_j z^{n_j}$ appearing in the classical Riesz product of remark **17.9**, its outer part has the constant term $\max\{\cos\theta_j, \sin\theta_j\}$ and the condition $\sum_{j=1}^{\infty}\cos^2\theta_j\sin^2\theta_j < \infty$ is equivalent to the condition $\prod_{j=1}^{\infty}\max\{\cos\theta_j, \sin\theta_j\}$ is positive. The additional information we have now is that in case μ is absolutely continuous with respect to dz, $\log\frac{d\mu}{dz}$ has finite integral.

2. It is to be noted that if each P_n is outer and the product $\prod_{k=1}^{\infty}P_k(0)$ is non-zero, then the formal expansion f of $\prod_{k=1}^{\infty}P_k(z)$ is an outer function. Indeed the Mahler measure of $\mid f \mid^2$ is $\mid f(0) \mid^2$, so f can not admit a non-trivial inner factor. Also the measure $1 \cdot dz$ can be expressed as a generalized Riesz product only by choosing each $P_n = 1$, for if any of the P_n is not the constant equal to 1, then its normalized outer part will have constant term less than one, which will force the Mahler measure of $1 \cdot dz$ to be less than 1, which is false. It is not known if the measure $cdz + d\delta_1, c, d > 0, c + d = 1$ can be expressed as a generalized Riesz product, where δ_1 denotes the Dirac measure at 1.

3. Let μ be a probability measure on S^1, and let q be a natural number. We contract the measure to the arc $A = \{z : z = \exp\{i\theta\}, 0 \le \theta < \frac{2\pi}{q}\}$, namely we consider the measure ν_1 supported on this arc given by $\nu_1(B) = \mu(z^q : z \in B), B \subset A$. We write similarly $\nu_j(C) = \nu_1(\exp\{-\frac{2\pi ij}{q}\}C), C \subset \exp\{\frac{2\pi ij}{q}\}A$. Let $\Pi_q(\mu) = \frac{1}{q}\sum_{j=1}^{q}\nu_j$. It can be verified that if $\mu = \mid P(z) \mid^2 dz$, then

$$\Pi_q(\mu) = \mid P(z^q) \mid^2 dz,$$

from which we conclude that if $\mu = \Pi_{k=1}^{\infty}\mid P_k(z) \mid^2$ then

$$\Pi_q(\mu) = \Pi_{k=1}^{\infty}\mid P_k(z^q) \mid^2 .$$

We see immediately that the Mahler measure of a generalized Riesz product is invariant under the application of Π_q for any q.

A Formula for Radon Nikodym Derivative

17.14. Consider two generalized Riesz products μ and ν based on polynomials $P_j, j = 1, 2, \cdots$ and $Q_j, j = 1, 2, \cdots$ where ν is continuous except for a possible mass at 1. Under suitable assumptions we prove the formula:

$$\sqrt{\frac{d\mu}{d\nu}} = \lim_{n\to\infty}\frac{\prod_{j=1}^{n}\mid P_j \mid}{\prod_{j=1}^{n}\mid Q_j \mid},$$

in the sense of $L^1(S^1, \nu)$ convergence.

Let σ and τ be two measures on the circle. Then, by Lebesgue decomposition of σ with respect to τ, we have

$$\sigma = \frac{d\sigma}{d\tau} d\tau + \sigma_s,$$

where σ_s is singular to τ and $\frac{d\sigma}{d\tau}$ is the Radon-Nikodym derivative. In the case of two Riesz products $\mu = \prod_{j=1}^{\infty} |P_j|^2$ and $\nu = \prod_{j=1}^{\infty} |Q_j|^2$, we are able to see that their affinities, namely the ratios $\frac{\prod_{j=1}^{n}|P_j|}{\prod_{j=1}^{n}|Q_j|}, n = 1, 2, \cdots$, converge in L^1 to $\sqrt{\frac{d\mu}{d\nu}}$, assuming that ν has no point masses except possibly at 1, This result extends a theorem of G. Brown and W. Moran [9]. Let δ_1 denote the unit mass at one. We have

17.15 Theorem. *Let $\mu = \prod_{j=0}^{\infty} |P_j|^2$, $\nu = \prod_{j=0}^{\infty} |Q_j|^2$ be two generalized Riesz products. Let*

$$\mu_n = \prod_{j=n+1}^{\infty} |P_j|^2, \quad \nu_n = \prod_{j=n+1}^{\infty} |Q_j|^2.$$

Assume that

1. $\nu = \nu' + b\delta_1$, ν' is continuous measure, $0 \le b < 1$.

2. $\prod_{j=0}^{n} |P_j|^2 d\nu_n \longrightarrow \mu$ weakly as $n \longrightarrow \infty$

3. $\prod_{j=0}^{n} |Q_j|^2 d\mu_n \longrightarrow \nu$ weakly as $n \longrightarrow \infty$.

Then the finite products $R_n = \prod_{k=1}^{n} \left| \frac{P_k(z)}{Q_k(z)} \right|, n = 1, 2, \cdots$ converge in $L^1(S^1, \nu)$ to $\sqrt{\frac{d\mu}{d\nu}}$.

To prove this we need the following proposition.

17.16. Proposition. *The sequence $\prod_{j=0}^{n} \left| \frac{P_j(z)}{Q_j(z)} \right|, n = 1, 2, \cdots$ converges weakly in $L^2(S^1, \nu)$ to $\sqrt{\frac{d\mu}{d\nu}}$.*

Proof. Put $f = \sqrt{\frac{d\mu}{d\nu}}$ and let n be a positive integer. Now

$$\int_{S^1} R_n^2 d\nu = \int_{S^1} \prod_{j=1}^{n} |P_j|^2 d\nu_n \to \int_{S^1} d\mu = 1$$

by assumption (2). Hence $\int_{S^1} R_n^2 d\nu, n = 1, 2, \cdots$ remain bounded. Thus, the weak closure of $R_n(z), n = 1, 2, \cdots$ in $L^2(S^1, \nu)$ is not empty.

We show that this weak closure has only one point, namely, $\sqrt{\frac{d\mu}{d\nu}}$. Indeed, let g be a weak subsequential limit, say, of $R_{n_j}(z), j = 1, 2, \cdots$. Then, for any continuous positive function h, we have, by judicious applications of Cauchy-Schwarz inequality,

$$\left(\int_{S^1} f(z)h(z)d\nu(z)\right)^2 = \left(\int_{S^1} h(z)R_{n_j}(z)\frac{1}{R_{n_j}}\sqrt{\frac{d\mu}{d\nu}}d\nu(z)\right)^2$$

$$\leq \left(\int_{S^1} h(z)R_{n_j}(z)d\nu(z)\right)\left(\int_{S_1} h(z)R_{n_j}(z)\frac{1}{R_{n_j}^2(z)}\frac{d\mu}{d\nu}d\nu(z)\right)$$

$$\leq \left(\int_{S^1} h(z)R_{n_j}(z)d\nu(z)\right)\left(\int_{S^1} h(z)\frac{1}{R_{n_j}(z)}d\mu\right)$$

$$\leq \int_{S^1} h(z)R_{n_j}(z)d\nu(z)\left(\int_{S^1} h(z)d\mu\right)^{\frac{1}{2}}\left(\int_{S^1} h(z)\frac{d\mu}{R_{n_j}^2(z)}\right)^{\frac{1}{2}}$$

$$\leq \left(\int_{S^1} h(z)R_{n_j}(z)d\nu(z)\right)\left(\int_{S^1} h(z)d\mu\right)^{\frac{1}{2}}\left(\int_{S^1} h(z)\mid\prod_{k=1}^{n_j} Q_k\mid^2 d\mu_{n_j}\right)^{\frac{1}{2}}.$$

Letting $j \to +\infty$, from our assumption (3), we get

$$\left(\int_{S^1} fhd\nu\right)^2 \leq \left(\int_{S^1} hgd\nu\right)\left(\int_{S^1} hd\mu\right)^{\frac{1}{2}}\left(\int_{S^1} hd\nu\right)^{\frac{1}{2}} \qquad (2)$$

But, since the space of continuous functions is dense in $L^2(\mu + \nu)$, we deduce from (2) that, for any Borel set B,

$$\left(\int_B fd\nu\right)^2 \leq \left(\int_B gd\nu\right)\left(\int_B d\mu\right)^{\frac{1}{2}}\left(\int_B d\nu\right)^{\frac{1}{2}}.$$

By taking a Borel set E such that $\mu_s(E) = 0$ and $\nu(E) = 1$, we thus get, for any $B \subset E$,

$$\left(\int_B fd\nu\right)^2 \leq \left(\int_B gd\nu\right)\left(\int_B f^2(z)d\nu\right)^{\frac{1}{2}}\left(\int_B d\nu\right)^{\frac{1}{2}}.$$

$$\left(\frac{1}{\nu(B)}\int_B fd\nu\right)^2 \leq \left(\frac{1}{\nu(B)}\int_B gd\nu\right)\left(\frac{1}{\nu(B)}\int_B f^2(z)d\nu\right)^{\frac{1}{2}}.$$

Since this holds for all Borel $B \subset E$, we conclude that for ν a.e. z, $(f(z))^2 \leq g(z)f(z)$, i.e., $f(z) \leq g(z)$ for ν a.e. z.. For the reverse inequality, note that for any continuous positive function h we have

$$\int_{S^1} ghd\nu = \lim_{j \to +\infty}\int_{S^1} h(z)R_{n_j}d\nu$$

$$\leq \lim_{j \to \infty} \left(\int_{S^1} h R_{n_j}^2 \, d\nu \right)^{\frac{1}{2}} \left(\int_{S^1} h d\nu \right)^{\frac{1}{2}}$$

$$\leq \left(\int_{S^1} h d\mu \right)^{\frac{1}{2}} \left(\int_{S^1} h d\nu \right)^{\frac{1}{2}}.$$

As before, for $B \subset E$,

$$\int_B g d\nu \leq \left(\int_B d\mu \right)^{\frac{1}{2}} \left(\int_B d\nu \right)^{\frac{1}{2}},$$

$$\int_B g d\nu \leq \left(\int_B f^2 d\nu \right)^{\frac{1}{2}} \left(\int_B d\nu \right)^{\frac{1}{2}},$$

and we deduce that $g(z) \leq f(z)$ for ν almost all z. So $g(z) = f(z)$, ν a.e.z. This completes the proof of the proposition.

Proof of Theorem 17.15. We will show that $\beta_n \overset{\text{def}}{=} \int_{S^1} \mid R_n - f \mid d\nu \to 0$ as $n \to \infty$, where $f = \sqrt{\frac{d\mu}{d\nu}}$. Now,

$$\frac{d\mu}{d\nu} = R_n^2(z) \frac{d\mu_n}{d\nu_n} \quad \text{and} \quad \sqrt{\frac{d\mu}{d\nu}} = R_n(z) \sqrt{\frac{d\mu_n}{d\nu_n}}.$$

Put

$$f_n^2 = \frac{d\mu_n}{d\nu_n},$$

then,

$$\int_{S^1} f_n^2 d\nu = \int_{S^1} \prod_{k=1}^n \mid Q_k \mid^2 d\mu_n \to \int_{S^1} d\nu = 1,$$

by assumption (3). The functions $f_n, n = 1, 2, \cdots$ are therefore bounded in $L^2(S^1, \nu)$. Hence, there exists a subsequence $f_{n_j} = \sqrt{\frac{d\mu_{n_j}}{d\nu_{n_j}}}, j = 1, 2, \cdots$ which converges weakly to some $L^2(S^1, \nu)$-function ϕ. We show that $0 \leq \phi \leq 1$ a.e (ν). For any continuous positive function h, we have

$$\left(\int_{S^1} h f_{n_j} d\nu \right)^2 \leq \left(\int_{S^1} h d\nu \right) \left(\int_{S^1} h f_{n_j}^2 d\nu \right)$$

$$\leq \left(\int_{S^1} h d\nu \right) \left(\int_{S^1} h \frac{d\mu_{n_j}}{d\nu_{n_j}} d\nu \right).$$

Hence, by letting j go to infinity combined with our assumption (3), we deduce that

$$\int_{S^1} h(z) \phi(z) d\nu \leq \int_{S^1} h(z) d\nu.$$

Since this hold for all continuous positive functions h, we conclude that $0 \leq \phi(z) \leq 1$ for almost all z with respect to ν. Thus any subsequential limit of the sequence $f_n, n = 1, 2, \cdots$ assumes values between 0 and 1. Now, for any subsequence $n_j, j = 1, 2, \cdots$ over which $f_{n_j}, j = 1, 2, \cdots$ has a weak limit , from our assumption (2) combined with Cauchy-Schwarz inequality, we have

$$\left(\int_{S^1} |R_{n_j} - f| d\nu \right)^2 = \left(\int_{S^1} |R_{n_j} - R_{n_j} f_{n_j}| d\nu \right)^2$$

$$= \left(\int_{S^1} R_{n_j} |1 - f_{n_j}| d\nu \right)^2$$

$$\leq \left(\int_{S^1} R_{n_j} |1 - f_{n_j}|^2 d\nu \right) \left(\int_{S^1} R_{n_j} d\nu \right)$$

$$\leq \left(\int_{S^1} R_{n_j} d\nu - 2 \int_{S^1} R_{n_j} f_{n_j} d\nu + \int_{S^1} R_{n_j} (f_{n_j})^2 d\nu \right) \left(\int_{S^1} R_{n_j} d\nu \right)$$

$$\leq \left(\int_{S^1} R_{n_j} d\nu - 2 \int_{S^1} f d\nu + \int_{S^1} R_{n_j} f_{n_j} \cdot f_{n_j} d\nu \right) \left(\int_{S^1} R_{n_j} d\nu \right)$$

$$\leq \left(\int_{S^1} R_{n_j} d\nu - 2 \int_{S^1} f d\nu + \int_{S^1} f \cdot f_{n_j} d\nu \right) \left(\int_{S^1} R_{n_j} d\nu \right).$$

Hence, letting j go to infinity,

$$\left(\lim_{j \to \infty} \int_{S^1} | R_{n_j} - f | d\nu \right)^2$$

$$\leq \int_{S^1} f d\nu - 2 \int_{S^1} f d\nu + \int_{S^1} f \cdot \phi d\nu$$

$$\leq \int_{S^1} (\phi(z) - 1) f(z) d\nu(z)$$

$$\leq 0,$$

and this implies that $R_{n_j}, j = 1, 2, \cdots$ converges to f in $L^1(S^1, \nu)$ and the proof of the theorem is achieved.

17.17 Remark. Notice that $\int_{S^1} \dfrac{d\mu}{d\nu} d\nu = 1$, implies the convergence of $\prod_{j=0}^{N} |R_j|$ to $\sqrt{\dfrac{d\mu}{dz}}$ in $L^2(d\nu)$.

We further have

17.18 Corollary. Two generalized Riesz products $\mu = \prod_{j=1}^{\infty} |P_j|^2$, $\nu = \prod_{j=1}^{\infty} |Q_j|^2$ satisfying the conditions of the above theorem are mutually singular if and only if

$$\int_{S^1} \prod_{j=0}^{n} \left| \frac{P_j}{Q_j} \right| d\nu \to 0 \text{ as } n \to \infty.$$

A Conditional Strong Dichotomy and Other Discussion

An important class of generalized Riesz products is the one arising in the study of rank one transformations of ergodic theory (see **15.28**).

17.19 Definition. A generalized Riesz product $\mu = \prod_{j=1}^{\infty} |P_j|^2$ is said to be of class (L) if for each sequence $k_1 < k_2 < \cdots$ of natural numbers, the tail measures $\prod_{j=n+1}^{\infty} |P_{k_j}|^2, n = 1, 2, \cdots$ converge weakly to Lebesgue measure.

17.20 Proposition. *If the generalized Riesz product* $\mu = \prod_{j=1}^{\infty} |P_j|^2$ *is of class (L) then the partial products* $\prod_{j=1}^{n} |P_j|, n = 1, 2, \cdots$ *converge in* $L^1(S^1, dz)$ *to* $\sqrt{\frac{d\mu}{dz}}$, *and the convergence is almost everywhere (w.r.t dz) over a subsequence.*

Proof. In Theorem **17.15** we put $Q_j(z) = 1$ for all j, so that ν is the Lebesgue measure on S^1. The first conclusion follows from theorem **17.15.**. The second conclusion follows since L^1 convergence implies convergence a.e over a subsequence.

The following formula follows immediately from this:

17.21 Corollary. *Let a generalized Riesz product* μ *be of class (L). Let* $\mathcal{K}_1, \mathcal{K}_2$ *be two disjoint subsets of natural numbers and let* $\mathcal{K}_0$ *be their union. Let* μ_1, μ_2 *and* μ_0 *be the generalized Riesz subproducts of* μ *over* $\mathcal{K}_1, \mathcal{K}_2$, *and* $\mathcal{K}_0$ *respectively. Then we have:*

$$\frac{d\mu_0}{dz} = \frac{d\mu_1}{dz} \frac{d\mu_2}{dz}, \tag{1}$$

where equality is a.e. with respect to the measure dz.

Let $\mu = \prod_{j=1}^{\infty} |P_j|^2$ be a generalized Riesz product of class (L). We assume that the polynomials $P_j, j = 1, 2, \cdots$ are outer. Write $S_n = \prod_{j=1}^{n} P_j$, and let $\phi_n = \frac{S_n}{|S_n|}$, a function of absolute value one. The functions $\phi_n, n = 1, 2, \cdots$ admit weak limits as functions in $L^{\infty}(S^1, dz)$. By Theorem **17.15** if β is positive then there is a unique nowhere vanishing weak limit $\frac{f}{|f|}$ which is indeed also a limit in $L^1(S^1, dz)$. On the other hand consider the simplest classical Riesz product given by

$$\mu = \prod_{j=1}^{\infty} \frac{1}{\sqrt{2}} \left| 1 + z^{n_j} \right|^2, \quad \frac{n_j}{n_{j-1}} \geq 3,$$

which is singular to the Lebesgue measure on S^1. Since $1 + e^{it} = |\,1 + e^{it}\,|\, e^{i\frac{t}{2}}$, we see that

$$\phi_k(e^{it}) = e^{i\left(\sum_{j=1}^{k} n_j\right)\frac{t}{2}} \longrightarrow 0$$

in the weak topology as $k \to \infty$, by virtue of the Riemann-Lebesgue lemma. However the following conditional strong dichotomy holds.

17.22 Theorem. *If the functions $\phi_n, n = 1, 2, \cdots$ admit a weak limit ϕ in $L^\infty(S^1, dz)$ which is non-vanishing a.e. (dz) on the set $\{z : \frac{d\mu}{dz} > 0\}$, then μ is either singular to Lebesgue measure, or its absolutely continuous part has positive Mahler measure.*

Proof. Let $f = \sqrt{\frac{d\mu}{dz}}$. Fix an integer k, then

$$\left| \int_{S^1} (z^k S_n(z) - z^k \phi_n(z) f(z)) dz \right| = \left| \int_{S^1} z^k \phi_n(z) (|S_n(z)| - f(z)) dz \right|$$

$$\leq \int_{S^1} ||S_n(z)| - f(z)|\, dz \to 0 \quad \text{as} \quad n \to \infty$$

by Theorem **17.15**. On the other hand, by assumption, since $f \in L^1(S^1, dz)$,

$$\int_{S^1} z^k \phi_n(z) f(z) dz \to \int_{S^1} z^k \phi(z) f(z) dz.$$

Now for $k < 0$, $\int_{S^1} z^k S_n(z) dz = 0$, so for $k < 0$,

$$\int_{S^1} z^k \phi(z) f(z) dz = 0.$$

By F and M Riesz theorem (K. H.Hoffman [12]) ϕf is either the zero function or a non-zero function in H^1. In the first case f is the zero function a.e dz, since ϕ is assumed to be non-vanishing a.e. (dz) on the set where f is positive. In the second case $|\,\phi f\,|$ has an integrable log, which implies that f has an integrable log. Thus f is either the zero function or has an integrable log. The theorem stands proved.

View the functions $S_n(z), n = 1, 2, \cdots$ as outer analytic functions on the open unit disk. From weak dichotomy theorem we immediately see that $S_n, n = 1, 2, \cdots$ converge uniformly on every compact subset of the open unit disk to a function which is non-zero and in H^1 if β is positive and the identically zero function if β is zero. We have, using notation from H^p theory, with $0 \leq r < 1$,

$$\lim_{r \to 1} \frac{1}{2\pi} \int_0^{2\pi} \left| S_n(re^{i\theta}) \right| d\theta = \frac{1}{2\pi} \int_0^{2\pi} \left| S_n(e^{i\theta}) \right| d\theta$$

We prefer to write this in our notation as

$$\lim_{r\to 1}\int_{S^1}|S_n(rz)|\,dz = \int_{S^1}|S_n(z)|\,dz$$

Letting $n \to \infty$ we get

$$\lim_{n\to\infty}\left(\lim_{r\to 1}\int_{S^1}|S_n(rz)|\,dz\right) = \lim_{n\to\infty}\left(\int_{S^1}|S_n(z)|\,dz\right) = \int\sqrt{\frac{d\mu}{dz}}\,dz.$$

However, in general one can not interchange the order of taking limits and write this as

$$= \lim_{r\to 1}\lim_{n\to\infty}\int_{S^1}|S_n(rz)|\,dz,$$

for that would immediately establish the singularity of μ with respect to the Lebesgue measure when $\beta = 0$, which is false in general, see remark **17.33**.

Non-Singular Rank One Maps and Generalized Riesz Products

17.23. In this section we will discuss generalized Riesz product in connection with spectral questions about non-singular and measure preserving rank one transformations. We will give necessary and sufficient conditions under which a generalized Riesz product is the maximal spectral type (up to possibly a discrete component) of a unitary operator associated with a rank one non-singular transformation and certain functions of absolute value one.

17.24 Non-Singular Rank One Maps. Let T be a non-singular rank one transformation obtained by cutting and stacking. This is done as follows. Let $\Omega_0 = \Omega_{0,0}$ denote the unit interval. At stage one of the construction we divide Ω_0 into m_1 pairwise disjoint intervals, $\Omega_{0,1}, \Omega_{1,1}\cdots,\Omega_{m_1-1,1}$, of lengths $p_{0,1}, p_{1,1},\cdots,p_{m_1-1,1}$, respectively, each $p_{i,j}$ being positive. Obviously $\sum_{j=0}^{m_1-1}p_{j,1}=1$. For each $j, 0 \le j \le m_1 - 2$, we stack $a_{j,1} \ge 0$ pairwise disjoint intervals of length $p_{j,1}$ on $\Omega_{j,1}$. Each interval is mapped linearly onto the one above it, except that $a_{j,1}$-th spacer is mapped linearly onto $\Omega_{j+1,1}$, $0 \le j \le m_1 - 2$. We thus get a stack of certain height h_1, together with a map T which is defined on all intervals of the stack except the interval at the top of the stack. Note that if $p_{j,1} \neq p_{j+1,1}$ for some j, T_1 will not be measure preserving. This completes the first stage of the construction.

At the k-th stage we divide the stack obtained at the the $(k-1)$-th stage in the ratios

$$p_{0,k}, p_{1,k},\cdots,p_{m_k-1,k}, \quad \sum_{i=0}^{m_k-1}p_{i,k}=1,$$

where each $p_{i,k}$ is positive. The spacers are added in the usual manner by which we mean that the spacers stacked above the j-th column are all of the same length which is the length of the top piece of the j-th column. The extension of T to the spacers is done linearly as usual. Note that the top of the spacers above the j-th column is mapped linearly onto the bottom of the $(j + 1)$-th column, so that if $p_{j,k} \neq p_{j+1,k}$, T will not be measure preserving. Note that at the k-th stage T is defined only on the algebra Γ_k generated by the levels of the k-th stack, except the top piece. The resulting T, after all the stages of the construction are completed, is defined on the space X consisting of increasing union of stack intervals. Let ν denote the Lebesgue measure defined on the σ-algebra Γ generated by $\cup_{n=1}^\infty \Gamma_n$. Note that $\prod_{j=1}^k p_{m_j-1,j}$ is the measure of the top piece of the column of height h_k at the end of kth stage of construction. We require that this goes to 0 as $k \to 0$. This ensures that T is eventually defined for almost every point of Ω_0. Note that T is non-singular (see remark below), ergodic with respect to ν, and $\frac{d\nu \circ T}{d\nu}$ is constant on all but the top layer of every stack. If no spacers are added at every stage of the construction, the resulting transformation will be called non-singular odometer.

17.25 Remark. The transformation T is non-singular in the sense that $m(T^{-1}(A)) = 0$ whenever $m(A) = 0$. For each j, let $p_{l_j,j} = \max\{\{p_{i,j} : 0 \leq i \leq m_j - 1\}$.

Consider the case when $\prod_{j=1}^\infty p_{l_j,j} > 0$. Then $\sum_{j=0}^\infty (1 - p_{l_j,j}) < \infty$. Now $\lambda_k \overset{def}{=} \prod_{j=1}^k p_{l_j,j}$ is the length of the largest of subinterval of $[0, 1)$ which appears as a level after the kth stage of construction. For each k, let I_k denote this level. Then $I_{k+1} \subset I_k$, the length of $I_{k+1} = \lambda_{k+1} = \lambda_k \times p_{l_{k+1},k+1}$ and $W \overset{def}{=} \cap_{k=1}^\infty I_k$ has positive length $= \prod_{j=1}^\infty p_{l_j,j}$. The Lebesgue measure of $(I_k - I_{k+1})$ is $\lambda_k - \lambda_{k+1}$. Write $L_k = (\cup_{j=-a_k}^{b_k} T^j(I_k - I_{k+1})) \cap [0, 1)$, where $T^{a_k}\Omega_{0,k} = I_k$, and $b_k = h_k - 1 - a_k$. Then the lebesgue measure of L_k is $1 - p_{l_{k+1},k+1}$, so by Borel-Canterlli Lemma the Lebesgue measure of $L \overset{def}{=} \limsup_{k\to\infty} L_k = \cap_{k=1}^\infty \cup_{j=k}^\infty L_j$ is zero. Now if $x \in [0, 1) - L$, then x is in at most finitely many L_js. This means that either $x \in W$ or for some fixed $y \in W$ and for some fixed integer $n(x)$, $x = T^{n(x)}y$. Thus we see that when $\prod_{j=1}^\infty p_{l_j,j}$ is non-zero, T induces a dissipative transformation on $[0, 1)$ which implies that T itself is dissipative in this case. There are two subcases: (i) if $l_j = 0$ for all j bigger than a fixed integer $N > 0$, then T is non-invertible and dissipative; W is the required wandering set which admits only finitely many negative iterates, but admits all positive iterates; (ii) if $l_j \neq 0$ for infinitely many j, then T is invertible and dissipative, W admits pairwise disjoint iterates over all integers. Note that the measure ν is defined on the σ-algebra generated by levels of the stacks so when $\prod_{j=1}^\infty p_{l_j,j} > 0$ we really have a discrete measure space, and ergodicity holds.

In case $\prod_{j=1}^\infty p_{l_j,j} = 0$, then T is defined on an atomfree measure space and the ergodicity of T follows from the usual Lebesgue density argument.

17.26 Unitary Operators U_T and V_ϕ. Let ϕ be a function on X of absolute value 1 which is constant on all but the top layer of every stack. On $L^2(X, \Gamma, \nu)$

define

$$(U_T f)(x) = \left(\frac{d\nu \circ T}{d\nu}(x) \right)^{1/2} f(Tx), f \in L^2(X, \Gamma, \nu)$$

$$(V_\phi f)(x) = (Vf)(x) = \phi(x) \cdot (U_T f)(x), f \in L^2(X, \Gamma, \nu).$$

U_T, and V are unitary operators, except when T is non-invertible, in which case U_T, V_ϕ are isometries isomorphic to the shift on l^2. The following argument is for the case when V is unitary, equivalently, when T is invertible.

$$(U_T^n f)(x) = \left(\frac{d\nu \circ T^n}{d\nu}(x) \right)^{1/2} f(T^n x),$$

$$(V^n f)(x) = \prod_{j=0}^{n-1} \phi(T^j(x)) \left(\frac{d\nu \circ T^n}{d\nu}(x) \right)^{1/2} f(T^n x).$$

It is mentioned in **15.35.** that the V has simple spectrum whose maximal spectral type (except possibly for some discrete part) is given by the generalized Riesz product

$$\prod_{j=1}^{\infty} p_{0,j} \mid P_j(z) \mid^2,$$

where

$$P_j(z) = 1 + c_{1,j} \left(\frac{p_{1,j}}{p_{0,j}} \right)^{1/2} z^{-R_{1,j}} + \cdots + c_{m_j-1,j} \left(\frac{p_{m_j-1,j}}{p_{0,j}} \right)^{1/2} z^{-R_{m_j-1,j}}.$$

The constants $c_{i,j}, 1 \le i \le m_j - 1, j = 1, 2, \cdots$, are of absolute value 1. They are determined by ϕ. The exponent $R_{i,j}, 1 \le i \le m_j - 1, j = 1, 2, \cdots$, is the i-th return time of a point in $\Omega_{0,j}$ into $\Omega_{0,j-1}$. It is equal to

$$R_{i,j} = ih_{j-1} + a_{0,j} + a_{1,j} + \cdots + a_{i-1,j}, 1 \le i \le m_j - 1.$$

We give the steps involved in proving this as it will allow us to make some needed observations. Write $Tf = f \circ T$. We have

$$(V^{-n} f)(\cdot) = T^{-n} \circ \left(\left(\prod_{j=0}^{n-1} \phi(T^j(\cdot)) \right)^{-1} \left(\frac{d(\nu \circ T^n)}{d\nu}(\cdot) \right)^{-1/2} f(\cdot) \right)$$

$$= \left(\prod_{j=0}^{n-1} \phi(T^{j-n}(\cdot)) \right)^{-1} \left(\frac{d\nu \circ T^n}{d\nu}(T^{-n}(\cdot)) \right)^{-1/2} f(T^{-n}(\cdot)),$$

whence

$$(T^{-n} f)(\cdot) = \prod_{j=0}^{n-1} \phi(T^{j-n}(\cdot)) \left(\frac{d(\nu \circ T^n)}{d\nu} \right)^{1/2} (T^{-n}(\cdot))(V^{-n} f)(\cdot).$$

Let $\Omega_{0,k-1}$ denote the base of the stack of height h_{k-1} after $(k-1)$-th stage of construction. Let $\Omega_{0,k}, \Omega_{1,k}, \cdots, \Omega_{m_k-1,k}$ be the partition of $\Omega_{0,k-1}$ during the k-th stage of construction, and let $a_{i,k}$ denote the number of spacers put on the column with base $\Omega_{i,k}$, $0 \le i < m_k - 1$. We have

$$\Omega_{0,k-1} = \cup_{j=0}^{m_k-1} T^{jh_{k-1}+\sum_{i=0}^{j-1} a_{i,k}}(\Omega_{0,k})$$

$$= \cup_{j=0}^{m_k-1} T^{R_{j,k}}(\Omega_{0,k}),$$

where

$$R_{j,k} = jh_{k-1} + \sum_{i=0}^{j-1} a_{i,k}$$

$=$ the j–th return time of a point in $\Omega_{0,k}$ into $\Omega_{0,k-1}$.

$$1_{\Omega_{0,k-1}} = \sum_{j=0}^{m_k-1} 1_{\Omega_{0,k}} \circ T^{-R_{j,k}},$$

$$= \sum_{j=0}^{m_k-1} c_{j,k} \left(\frac{d\nu \circ T^{R_{j,k}}}{d\nu} (T^{-R_{j,k}}) \right)^{1/2} (\cdot)(V^{-R_{j,k}} 1_{\Omega_{0,k}})(\cdot),$$

where $c_{j,k} = \prod_{i=0}^{R_{j,k}-1} \phi(T^{i-R_{j,k}}(\cdot))$, a constant of absolute value one. Note that the constants $c_{j,k}$'s can be preassigned and ϕ can be so defined that the above relation holds for all (j,k). We now observe that for $x \notin T^{R_{j,k}}\Omega_{0,k}$,

$$V^{-R_{j,k}} 1_{\Omega_{0,k}}(x) = 0,$$

and that for $x \in T^{R_{j,k}}\Omega_{0,k}$,

$$\frac{d\nu \circ T^{R_{j,k}}}{d\nu}(T^{-R_{j,k}}(x)) = \frac{p_{j,k}}{p_{0,k}}.$$

We thus have

$$1_{\Omega_{0,k-1}} = \sum_{j=0}^{m_k-1} c_{j,k} \left(\frac{p_{j,k}}{p_{0,k}} \right)^{1/2} (V^{-R_{j,k}} 1_{\Omega_{0,k}}).$$

Let us normalize $1_{\Omega_{0,k}}$ and write

$$f_k = \left(\frac{1}{m(\Omega_{0,k})} \right)^{1/2} 1_{\Omega_{0,k}} = \left(\frac{1}{(\prod_{j=1}^k p_{0,j})} \right)^{1/2} 1_{\Omega_{0,k}},$$

$$f_{k-1} = (p_{0,k})^{1/2} \left(\sum_{j=0}^{m_k-1} c_{j,k} \left(\frac{p_{j,k}}{p_{0,k}} \right)^{1/2} V^{-R_{j,k}} \right) 1_{\Omega_{0,k}}.$$

Now $m(\Omega_{0,0}) = 1$ so $f_0 = 1_{\Omega_{0,0}}$. We have by iteration

$$f_0 = \left(\prod_{j=1}^{k} P_j(V) \right) f_k,$$

where

$$P_j(z) = (p_{0,j})^{1/2} (1 + c_{1,j} (\frac{p_{1,j}}{p_{0,j}})^{1/2} z^{-R_{1,j}} + \cdots + c_{m_j-1,j} (\frac{p_{m_j-1,j}}{p_{0,j}})^{1/2} z^{-R_{m_j-1,j}}).$$

Let $V^n = \int_{S^1} z^{-n} dE, n \in \mathbb{Z}$, be the spectral resolution of the unitary group $V^n, n \in \mathbb{Z}$, and let

$$(V^n f_k, f_k) = \int_{S^1} z^{-n} (E(dz) f_k, f_k) = \int_{S^1} z^{-n} d\sigma_k$$

where $\sigma_k(\cdot) = (E(\cdot) f_k, f_k)$.

We therefore have for all integers l

$$(V^l f_0, f_0) = \int_{S^1} z^{-l} d\sigma_0 = \int_{S^1} z^{-l} \prod_{j=0}^{k} |P_j(z)|^2 d\sigma_k,$$

whence

$$d\sigma_0 = \prod_{j=1}^{k} |P_j(z)|^2 d\sigma_k.$$

Now we will show, as in the measure preserving case (see **15.29**), that σ_0 is the generalized Riesz product:

$$\sigma_0 = \prod_{j=1}^{\infty} |P_j(z)|^2 .$$

Let N_k denote the the set of integers consisting of zero together with the entry times of a point in $\Omega_{0,k}$ into $\Omega_{0,0}$ which are less than the height h_k of the kth stack.

We have

$$f_0 = \left(\prod_{j=1}^{k} P_j(V) \right) f_k = Q_k(V) f_k,$$

where

$$Q_k(z) = \prod_{j=1}^{k} P_j(z) \stackrel{def}{=} \sum_{j=0}^{h_k-1} q_j(k) z^j = \sum_{j \in N_k} q_j(k) z^j$$

an expansion of the product of dissociated polynomials $P_1, P_2, \cdots, P_k$. Note that

(i) $\mid q_j \mid \leq 1$

(ii) $\mid q_r \mid \leq \prod_{j=1}^{k-1} p_{m_j-1,j} \to 0$ as $k \to \infty$, for $h_k - h_{k-1} < r < h_k$

$$\int_{S^1} z^n \mid Q_k \mid^2 dz = \sum_{r-s+n=0} q_r(k)\overline{q_s(k)}$$

where $r, s \in N_k$.

Now fix $n \in \mathbb{Z}$ and let k be so large that the first return time for any $x \in \Omega_{0,k}$ back to $\Omega_{0,k}$ is bigger than $\mid n \mid$, i.e, k is so large that $h_k \geq \mid n \mid$. We actually choose k so large that $\mid n \mid < \frac{h_{k-1}}{2}$. If $r, s \in N_k$ then $r - s + n$ can never exceed or equal the second return of an $x \in \Omega_{0,k}$ back to $\Omega_{0,k}$ (under T or T^{-1}). Moreover there can be at most n^2 pairs (r, s) with $r, s \in N_k$ with $T^{r+n-s}\Omega_{0,k} \cap \Omega_{0,k} \neq \emptyset$. For suppose $n > 0$ and $T^{r+n-s}\Omega_{0,k} \cap \Omega_{0,k} \neq \emptyset$ and $r + n - s \neq 0$, $r, s \in N_k$. Then $r + n - s = u$ where u is the first return time of a point $x \in \Omega_{0,k}$ back to $\Omega_{0,k} \geq h_k$. $s = r + n - u$. Since n, r, s are less than h_k, $h_k \leq u$ and $s \geq 0$, we have $0 \leq s < n$ and $n - s + r = u \geq h_k$, so $r \geq h_k - (n-s) \geq h_k - n$. Thus there can be at most n^2 pairs (r, s) with $r, s \in N_k$ such that $T^{n+r-s}\Omega_{0,k} \cap \Omega_{0,k} \neq \emptyset$. Thus if $T^{n+r-s}\Omega_{0,k} \cap \Omega_{0,k} \neq \emptyset$, $r, s \in N_k$ then $n + r - s = 0$ except for at most n^2 pairs (r, s), $r, s \in N_k$. This in turn implies that $(V^{n+r-s}1_{\Omega_{0,k}}, 1_{\Omega_{0,k}}) = 0$ except when $n + r - s = 0$ and at most n^2 other pair (r,s), $r, s \in N_k$.

$$(V^n f_0, f_0) = (V^n Q_k(V)f_k, Q_k(V)f_k) = (V^n \mid Q_k \mid^2 (V)f_k, f_k)$$

$$= \sum_{n+r-s=0, r,s \in N_k} q_r \overline{q_s}(V^{n+r-s}f_k, f_k) + \sum_1$$

$$= \sum_{r-s+n=0, r,s \in N_k} q_r(k)\overline{q_s(k)} + \sum_1,$$

where $\sum_1$ is a sum of at most n^2 terms of the type

$$q_r \overline{q_s}(V^{n+r-s}f_k, f_k), n + r - s \neq 0.$$

Now $h_k - h_{k-1} < h_k - n \leq r \leq h_k - 1$, so that $\mid q_r(k) \mid \leq \prod_{j=1}^{k-1} p_{m_j-1,j} \to 0$ as $k \to \infty$. Clearly then the sum $\sum_1$ goes to zero as $k \to \infty$ and the claim is proved.

17.27. Remark. If $\prod_{j=1}^{\infty} p_{l_j,j} = 0$ and $V_\phi = U_T$, then it can be verified that $\sum_{k=1}^{\infty} \mid \hat{\sigma}_0(k) \mid^2 \geq \sum_{j=1}^{\infty}(1 - p_{l_j,j}) = \infty$. In addition if for some M, $m_k \leq M < \infty$ for all k, then σ_0 is singular to Lebesgue measure.

Generalized Riesz Products of Dynamical Origin

Consider now the polynomials appearing in the above generalized Riesz product.

$$P_j(z) = (p_{0,j})^{1/2}(1 + c_{1,j}(\frac{p_{1,j}}{p_{0,j}})^{1/2}z^{-R_{1,j}} + \cdots + c_{m_j-1,j}(\frac{p_{m_j-1,j}}{p_{0,j}})^{1/2}z^{-R_{m_j-1,j}})$$

The exponent $R_{i,j}, 1 \leq i \leq m_j - 1, j = 1, 2, \cdots$, is the i-th return time of a point in $\Omega_{0,j}$ into $\Omega_{0,j-1}$. Also

$$R_{i,j} = ih_{j-1} + a_{0,j} + a_{1,j} + \cdots + a_{i-1,j}, 1 \leq i \leq m_j - 1$$

where h_{j-1} is the height of the tower after $(j-1)$-th stage of the construction is complete, and $a_{k,j}$ is the number of spacers on the k-th column, $0 \leq k \leq m_j - 2$. We observe that

1. $h_1 = R_{m_1-1,1} + 1$,

2. $R_{1,j} \geq h_{j-1} > R_{m_{j-1},j-1}$,

3. $R_{i+1,j} - R_{i,j} \geq h_{j-1}$.

These properties (1), (2), (3) of the powers $R_{i,j}$, $1 \leq i \leq m_j - 1, j = 1, 2, \cdots$ indeed characterize generalized Riesz products which arise from nonsingular rank one transformations (together with a ϕ) in the above fashion. More precisely consider a generalized Riesz product

$$\prod_{j=1}^{\infty} |Q_j(z)|^2 .$$

where

$$Q_j(z) = \sum_{i=0}^{n_j} b_{i,j} z^{r_{i,j}}, \quad b_{i,j} \neq 0, \quad \sum_{i=0}^{n_j} |b_{i,j}|^2 = 1, \prod_{j=1}^{\infty} |b_{n_j,j}| = 0.$$

Define inductively:

$$h_0 = 1, h_1 = r_{n_1,1} + h_0, \cdots, h_j = r_{n_j,j} + h_{j-1}, j \geq 2$$

Note that $h_j > r_{n_j,j}$.

17.28 Proposition. *Assume that for each $j = 1, 2, \cdots$,*

$$r_{1,j} \geq h_{j-1}, \quad r_{i+1,j} - r_{i,j} \geq h_{j-1}$$

Then $r_{i,j}, h_j$, satisfy (1), (2) and (3). The generalized product $\prod_{j=1}^{\infty} |Q|^2$ describes the maximal spectral type (up to possibly a discrete part) of a suitable V_ϕ.

Proof. That the $r_{i,j}, h_j$ satisfy (1), (2), (3) is obvious. The needed non-singular T is given by cutting parameters $p_{i,j} = |b_{i,j}|^2, i = 0, 1, \cdots, n_j, j = 1, 2, \cdots$, and spacers $a_{i-1,j} = r_{i,j} - r_{i-1,j} - h_{j-1}, 1 \leq i \leq n_j, j = 1, 2, \cdots$. The needed ϕ (which need not be unique) is given by constants $\frac{b_{i,j}}{|b_{i,j}|}, 0 \leq i \leq n_j, j = 1, 2, \cdots$. This proves the proposition.

16.29 Definition. A generalized Riesz product $\mu = \prod_{j=1}^{\infty} \mid Q_j(z) \mid^2$, where $Q_j(z) = \sum_{i=0}^{n_j} b_{i,j} z^{r_{i,j}}, b_{i,j} \neq 0, \sum_{i=0}^{n_j} \mid b_{i,j} \mid^2 = 1, \prod_{j=1}^{\infty} \mid b_{n_j,j} \mid = 0$, is said to be of dynamical origin if with

$$h_0 = 1, h_1 = r_{n_1,1} + h_0, \cdots, h_j = r_{n_j,j} + h_{j-1}, j \geq 2$$

it is true that for $j = 1, 2, \cdots,$

$$r_{1,j} \geq h_{j-1}, \quad r_{i+1,j} - r_{i,j} \geq h_{j-1}$$

If, in addition, the coefficients $b_{i,j}$ are all positive, then we say that μ is of purely dynamical origin.

Flat Polynomials, Generalized Riesz Products, Banach's Problem

17.30 Lemma *Given a sequence* $P_n = \sum_{j=0}^{m_n} a_{j,n} z^j, , n = 1, 2, \cdots$ *of analytic trigonometric polynomials in* $L^2(S^1, dz)$ *with non-zero constant terms and* $L^2(S^1, dz)$ *norm 1,* $\prod_{n=1}^{\infty} \mid a_{m_n,n} \mid = 0$, *there exist a sequence of positive integers* $N_1, N_2, \cdots$ *such that*

$$\prod_{j=1}^{\infty} \mid P_j(z^{N_j}) \mid^2$$

is a generalized Riesz product of dynamical origin.

Proof For each $j \geq 1$, by dropping terms of P_j with coefficient zero, we can write P_j as:

$$P_j = \sum_{i=0}^{n_j} b_{i,j} z^{r_{i,j}}, b_{i,j} \neq 0, \quad b_{0,j} > 0, \quad \sum_{i=1}^{n_j} \mid b_{i,j} \mid^2 = 1.$$

Let $N_1 = 1$ and $h_1 = H_1 = r_{n_1,1} + 1$. Choose $N_2 \geq 2H_1 > 2r_{n_1,1}$. Then

$$N_2 \cdot r_{1,2} > h_1, N_2(r_{i+1,2} - r_{i,2}) > h_1.$$

Since $N_2 > 2r_{n_1,1}$ the polynomials $\mid P_1(z^{N_1}) \mid^2$ and $\mid P_2(z^{N_2}) \mid^2$ are dissociated. Consider now $P_1(z^{N_1})P_2(z^{N_2})$. Write $H_2 = N_1 r_{n_1,1} + N_2 r_{n_2,2} + h_1 > N_2 r_{n_2,2} + h_1 \overset{def}{=} h_2$. Choose $N_3 \geq 2H_2$. Then

$$N_3 \cdot r_{1,3} \geq h_2, N_3(r_{i+1,3} - r_{i,3}) > h_2.$$

Since $N_3 \geq 2H_2 > 2(N_1 r_{(n_1,1)} + N_2 r_{(n_2,2)})$ the polynomial $\mid P_3(z^{N_3}) \mid^2$ is dissociated from $\mid P_1(z^{N_1})P_2(z^{N_2}) \mid^2$. Proceeding thus we get $N_j, j = 1, 2, \cdots$ and polynomials $Q_j(z) = P_j(z^{N_j}), j = 1, 2, \cdots$ such that
(i) $\parallel Q_j \parallel_2 = 1$ (since $\parallel P_j \parallel_2 = 1$ and the map $z \rightarrow z^{N_j}$ is measure preserving.) (ii) the polynomials $\mid Q_j \mid^2, j = 1, 2, \cdots$ are dissociated, (iii) for each

$j \geq 1$,

$$h_{j-1} < N_j r_{1,j}, \quad h_{j-1} < N_j (r_{i+1,j} - r_{i,j})$$

Since the polynomials $Q_j, j = 1, 2, \cdots$ have $L^2(S^1, dz)$ norm 1 and their absolute squares are dissociated, the generalized Riesz product $\prod_{j=1}^{\infty} |P(z^{N_j})|^2$ is well defined. Moreover, (iii) shows that the conditions for it to arise from a non-singular rank one T and a ϕ in the above fashion are satisfied. The lemma follows.

An immediate application of this Lemma is the following:

17.31 Theorem *Let* $P_j = \sum_{i=0}^{m_j} a_{i,j} z^i, j = 1, 2, \cdots$ *be a sequence of analytic trigonometric polynomials with unit* $L^2(S^1, dz)$-*norm, satisfying for all* j, $|a_{m_j,j}| < \lambda < 1$, *and such that* $|P_j(z)| \to 1$ *a.e.* (dz) *as* $j \to \infty$. *Then there exists a subsequence* $P_{j_k}, k = 1, 2, \cdots$ *and natural numbers* $N_1 < N_2 < \cdots$ *such that the product* $\mu = \prod_{k=1}^{\infty} |P_{j_k}(z^{N_k})|^2$ *is a generalized Riesz product of dynamical origin with* $\frac{d\mu}{dz} > 0$ *a.e.* (dz).

Proof. Since $|P_j(z)| \to 1$ as $j \to \infty$ a.e. (dz), by Egorov's theorem we can extract a subsequence $P_{j_k}, k = 1, 2, \cdots$ such that the sets

$$E_k \stackrel{def}{=} \left\{ z : |(1 - |P_{j_l}(z)|)| < \frac{1}{2^k} \ \forall \ l \geq k \right\}$$

increase to S^1 (except for a dz null set), and $\sum_{k=1}^{\infty} (1 - dz(E_k)) < \infty$. Write $Q_k = P_{j_k}$. Then for $z \in E_k$, $|(1 - |Q_k(z)|)| < \frac{1}{2^k}$. Note that $\prod_{k=1}^{\infty} |a_{m_{j_k},j_k}| = 0$ since for all k, $|a_{m_{j_k},j_k}| < \lambda < 1$. By lemma **17.30** we can choose $N_1, N_2, \cdots$ such that

$$\prod_{k=1}^{\infty} |Q_k(z^{N_k})|^2$$

is a generalized Riesz product of dynamical origin . We show that $\lim_{L \to \infty} \prod_{k=1}^{L} |Q_k(z^{N_k})|$ is nonzero a.e. (dz), which will imply, by proposition **17.20**, that $\frac{d\mu}{dz} > 0$ a.e (dz).

Now the maps $S_k : z \to z^k, k = 1, 2, \cdots$ preserve the measure (dz), and since $\sum_{k=1}^{\infty} dz(S^1 - E_k) < \infty$ we have $\sum_{k=1}^{\infty} dz(S_{N_k}^{-1}(S^1 - E_k)) < \infty$. Let $F_k = S_{N_k}^{-1}(S^1 - E_k)$ and $F = \limsup_{k \to \infty} F_k = \cap_{k=1}^{\infty} \cup_{l=k}^{\infty} F_l$. Then $dz(F) = 0$, and if $z \notin F$, $z \notin S_{N_k}^{-1}(S^1 - E_k)$ hold for all but finitely many k, which in turn implies that $S_{N_k} z \in E_k$ for all but finitely many k. Thus, if $z \notin F$, then $|(1 - |Q_k(z^{N_k})|)| < \frac{1}{2^k}$ for all but finitely many k. Also the set of points z for which some finite product $\prod_{k=1}^{L} |Q_k(z^{N_k})|$ vanishes is countable. Clearly $\lim_{L \to \infty} \prod_{k=1}^{L} |Q_k(z^{N_k})|$ is nonzero a.e. (dz) and the theorem is proved.

17.32 Corollary *(i) If $P_k, k = 1, 2, \cdots$ are as in the above theorem and if* $\limsup_{k \to \infty} M(P_k) = 1$, *then we can choose* $P_{j_k}, k = 1, 2, \cdots$ *and* $N_1, N_2, \cdots$, *in such a way that* $M(\mu)$ *is positive.*
(ii) If $P_k, k = 1, 2, \cdots$ are as in the above theorem and if $\liminf_{k \to \infty} M(P_k) < 1$ *then we can choose* $P_{j_k}, k = 1, 2, \cdots$ *and* $N_1, N_2, \cdots$ *in such a way that* $M(\mu) = 0$, *and* $\frac{d\mu}{dz} > 0$ *a.e. (dz).*

17.33 Remark. Now it is easy to construct polynomials $P_k, k = 1, 2, \cdots$ satisfying the hypothesis of part (ii) of the above corollary. For example one can modify the flat sequence of polynomials given in **17.41** so that the modified sequence remains flat, the coefficients of polynomials therein remain uniformly away from 1, and their Mahler measures converge to 0. So one can obtain generalized Riesz products μ with zero Mahler measure and $\frac{d\mu}{dz}$ positive a.e (dz). Thus the interchange of order of limits discussed after the proof of theorem **17.22** is not always possible without some additional conditions.

17.34. A unitary operator U on a complex separable Hilbert space H is said to have simple Lebesgue spectrum if it has multiplicity one and its maximal spectral type is the Lebesgue measure on the circle, equivalently, there is a vector f_0 in H such that $U^n f_0, n \in \mathbb{Z}$, form a complete orthonormal basis in H.

We discuss in the rest of this section the problem of 'simple Lebesgue spectrum' in ergodic theory in the light of theorem **17.31**.

We say that U has simple Lebesgue component or Lebesgue component of multiplicity one in case we can decompose H into U invariant orthogonal closed subspaces H_1, H_2 such that U restricted to H_1 has simple Lebesgue spectrum and maximal spectral type of U restricted to H_2 is singular to Lehesgue measure.

It is well known that a two sided Bernoulli shift has countably infinite Lebesgue spectrum on the ortho-complement of constant functions, i.e., for a two sided Bernoulli shift T on $[0, 1]$ the ortho-complement of constants in $L^2([0, 1])$ can be decomposed into countably infinite number of U_T invariant closed subspaces on each of which U_T has simple Lebesgue spectrum.

The only known measure preserving transformation T on a measure space such that $U_T : f \to f \circ T, f \in L^2$, has simple Lebesgue spectrum is the one which is isomorphic to the translation by one on the integers.

17.35. A problem, attributed to Banach, asks whether there exists an invertible Lebesgue measure preserving transformation T on the real line such that

$$U_T : f \to f \circ T, f \in L^2(R),$$

has simple Lebesgue spectrum, (S. Ulam [16], p76).

This question has many variants.

Does there exist a Lebesgue measure preserving transformation T on $[0,1]$ such that U_T has simple Lebesgue spectrum on the ortho-complement of constant functions.

We can ask more generally if there exists an invertible non-singular T on the real line such that $U_T : f \rightarrow \sqrt{\frac{dL \circ T}{dx}} f \circ T, f \in L^2(R)$, has simple Lebesgue spectrum. (L denotes the Lebesgue measure on R.)

One can ask similar question for measure preserving flows.

All three versions of the problem remain open, at least until the presently available solutions (E. Abdalaoui [2], A. Prikhodko [15]) become tractable.

We ask a weaker question: Does there exist an invertible measure preserving T on $[0,1]$ or the real line such that U_T has a Lebesgue *component* of multiplicity one?

We connect this weaker version to the existence of suitable sequence of flat polynomials.

17.36 Definition. A sequence $P_n, n = 1, 2, \cdots$ of analytic trigonometric polynomials, each P_n of $L^2(S^1, dz)$ norm 1, is said to be a.e flat if $|P_n(z)| \rightarrow 1$ for a.e. $z \in S^1$.

Consider the Bourgain class of (B) of all polynomials of the type

$$P(z) = \frac{1}{\sqrt{m+1}}(1 + z^{n_1} + z^{n_2} + \cdots + z^{n_m}),$$

where $0 < n_1 < n_2 < \cdots < n_m$. Since $L^2(S^1, dz)$ norm of such a P is one, its $L^1(S^1, dz)$-norm is at most one. J. Bourgain [7] has raised the question if the supremum of the $L^1(S^1, dz)$-norms of elements in (B) is 1. If the answer to this question is affirmative, then there is a sequence $P_n, n = 1, 2, \cdots$, of elements in (B) which is a.e. flat. By theorem **17.31** there is a subsequence $P_{j_n}, n = 1, 2, \cdots$ and increasing positive integers $N_n, n = 1, 2, \cdots$ such that the generalized Riesz product $\mu = \Pi_{n=1}^\infty |P_{j_n}|^2$ is of dynamical origin and $\frac{d\mu}{dz} > 0$ a.e. Since the coefficients of $P_{j_n}, n = 1, 2, \cdots$ are all positive, μ is of purely dynamical origin. We conclude:

17.37. *If supremum*$\{|| P ||_1 : P \in B\} = 1$, *then there is a measure preserving rank one transformation* T *such that spectrum of* U_T *admits a Lebesgue component of multiplicity one.*

Similarly we have:

17.38. *If there is an a.e. flat sequence* $P_n = \sum_{j=0}^{m_n} a_{j,n} z^j, || P_n ||_2 = 1, n = 1, 2, \cdots$ *with coefficients* $a_{j,n}, 0 \leq j \leq m_n, n = 1, 2, \cdots$ *non-negative and uniformly bounded away from 1, then there exists a nonsingular rank one* T *such that spectrum of* U_T *admits a simple Lebesgue component.*

Consider the class U of polynomials of the type $\frac{1}{\sqrt{n}} \sum_{j=0}^{n} a_j z^j, a_j \in \{-1, +1\}, 0 \le j \le n$. It is not known if the supremum of $L^1(S^1, dz)$ norms of polynomials in U is one. M. Guenais has shown that if this supremum is one then there is a finite measure preserving transformation T such that U_T has a Lebesgue component of multiplicity one. In the light of theorem **17.31** and **5.6** this can be proved as follows. Since $\sup\{\| P \|_1 : P \in U\} = 1$ there is a sequence $P_j.j = 1, 2, \cdots$, of elements in U which converges a.e. in absolute value to 1. By theorem **17.31.** there is a generalized Riesz product $\mu = \Pi_{k=1}^{\infty} | P_{j_k}(z^{N_k}) |^2$ of dynamical origin such that $\frac{d\mu}{dz} > 0$ a.e. So, there is a non-singular rank one τ and a ϕ of absolute value one such that $V_\phi = \phi U_\tau$ admits a Lebesgue component of multiplicity one. Since P_ns are from the class U, the τ turns out to be a general measure preserving odometer, while ϕ assumes values -1 and $+1$. If we write T for the skew product transformation on $X \times \{-1, 1\}$

$$T(x, y) = (\tau x, \phi(x) y),$$

then by **5.6.** U_T admits a Lebesgue component of multiplicity one since V_ϕ does. We have proved the following result due to M. Guenais: [11]:

17.39. *If the supremum of $L^1(S^1, dz)$ norms of elements in U is one then there is a measure preserving T such that spectrum of U_T admits a Lebesgue component of multiplicity one.*

A finite sequence $(e_0, e_2, \cdots, e_{n-1})$ of $+1$ and -1 is said to be a Barker sequence if for all $k, 0 < k \le n - 1$, the aperiodic correlation

$$\sum_{j=0}^{n-k} e_j e_{j+k}$$

does not exceed 1 in absolute value. It is known that there are only finitely many Barker sequences of odd length, and there are no Barker sequences of odd length greater than 13. It is not known if there are infinitely many Barker sequences, and it is conjectured that there are only finitely many Barker sequences. P. Borwein and M. Mossinghoff [6] have shown that if $e_0, e_1, \cdots, e_{n-1}$ is a Barker sequence of length n and if

$$P(z) = \frac{\sum_{j=0}^{n-1} e_j z^j}{\sqrt{n}},$$

then the Mahler measure $M(P)$ of P is $> 1 - \frac{1}{n}$. This immediately implies that the supremum of $L^1(S^1, dz)$ norms of such polynomials is 1. As before we can prove the following result due to T. Downarowich, Y. Lacroix[10] :

17.40. *If there are infinitely many Barker sequences then there is a measure preserving T such that spectrum of U_T has Lebesgue component of multiplicity one.*

17.41. The existence of transformations in **17.37 - 17.40** is speculative: they exist if flat sequences polynomials of certain kind exist. If we do not insist that the transformation be measure preserving, but require only that it be a non-singular transformation on an atom free measure space, then we can give a concrete example of a T such that U_T has Lebesgue component of multiplicity one. Consider the power series expansion of the linear fractional map $\frac{z-\alpha}{1-\alpha z}$, $0 < \alpha < 1$, which maps S^1 onto itself:

$$\frac{z-\alpha}{1-\alpha z} = -\alpha + \sum_{j=1}^{\infty}(\alpha^{j-1} - \alpha^{j+1})z^j.$$

Let $P_n = -\alpha + \sum_{j=1}^{n}(\alpha^{j-1} - \alpha^{j+1})z^j$. Then the sequence $\frac{P_n}{||P_n||_2}, n = 1, 2, \cdots$ converges in absolute value uniformly to 1. As in **17.39.** we can now get a non-singular T such that U_T has Lebesgue component of multiplicity one in its spectrum. (see El. Abdalaoui and M. Nadkarni [5])

17.42. Conditions for singularity of generalized Riesz products made of polynomials from Bourgain class (B) are well discussed in the literature on the subject. We mentions some of the papers on this. J. Bourgain [7] has shown that for a.e. $\omega \in \Omega$ the Ornstein transformation T_ω defined in **16.11** has singular spectrum. Klemes and Reinhold [13] have shown that if $\sum_{k=1}^{\infty} \frac{1}{m_k^2} = \infty$ where m_k the degree of the the kth p0lynomial of a Riesz product $\mu = \Pi_{k=1}^{\infty} \mid P_k \mid^2$, $P_k, \in B, k = 1, 2, \cdots$, then μ is singular to Lebesgue measure. El Abdalaoui [1] has given conditions for singularity of μ in case $\sum_{k=1}^{\infty} \frac{1}{m_k^2} < \infty$. In [15] A. Prikhodko discusses connection of flat polynomials with Banach's problem for flows. In papers [2] and [3] E. Abdalaoui has discussed flat polynomials in connection with Banach's problem and Barker sequences respectively, in addition, these papers provide a fairly exhaustive list of papers on flat polynomials and related topics.

Zeros of Polynomials

17.43. Consider the polynomial of the type

$$P(z) = 1 + z^{h+a_1} + z^{2h+a_1+a_2} + \cdots + z^{(m-1)h+a_1+a_2+\cdots+a_{m-1}}, \qquad (R)$$

which appears in the generalized Riesz product connected with rank one measure preserving transformation. It is easy to see that zeros of these polynomials cluster near the unit circle as h tends to ∞. We prove a quantitative result, namely, if w is a zero of this polynomial then

$$\left(\frac{1}{2}\right)^{\frac{1}{h}} \leq \mid w \mid \leq (2)^{\frac{1}{h}}$$

To see this we write $\mid w \mid = a$. Assume first that $a \leq 1$. Then, since w is a zero of P,

$$a^h + a^{2h} + \cdots + a^{(m-1)h} \geq 1.$$

Equivalently,

$$a^h \frac{(1 - a^{(m-1)h})}{1 - a^h} \geq 1.$$

$$a^h - a^{mh} \geq 1 - a^h$$

$$2a^h \geq 1 + a^{mh} \geq 1$$

which proves the result when $|w| \leq 1$. To prove the second half of the inequality we note that if $\mid w \mid$ is greater than 1 then $\frac{1}{|w|} < 1$ and $\frac{1}{w}$ is a zero of $P(\frac{1}{z})$ so the second half follows from the first half. A slight improvement of the inequality is possible. If $m = 2$ then all the zeros of P lie on the unit circle. It is easy to show that if $m > 2$ then the equation $x^m - 2x + 1$ has a zero, say b_m, in the open interval $\frac{1}{2} < x < 1$. and one can show that

$$(b_m)^{\frac{1}{h}} \leq \mid w \mid \leq \left(\frac{1}{b_m}\right)^{\frac{1}{h}}.$$

However, it is not a very big improvement since one can show that $b_m \to \frac{1}{2}$ as $m \to \infty$. This simple result tells us that if each P_k has less than ch_{k-1} zeros bigger than 1 in absolute value, where c is a positive constant less than one, then $\prod_{k=1}^{\infty} |\alpha_k| = 0$.

We mention that A. Odlyzko and B. Poonen [14] have proved that the zeros of the polynomials with coefficients in $\{0, 1\}$ are contained in the annulus $\frac{1}{\phi} < |z| < \phi$ where ϕ is the golden ratio.

Zeros of polynomials with restricted coefficients has deep and extensive literature. We state here only a result by P. Brown, T. Erdélyi, F. Littmann [8]. Let

$$K_n = \left\{ \sum_{k=0}^{n} a_k z^k : \mid a_0 \mid = \mid a_n \mid = 1, \mid a_k \mid \leq 1 \right\},$$

and let n be so large that $\delta_n = 33\pi \frac{log(n)}{\sqrt{n}} < 1$, then any polynomial in K_n admits at least $8\sqrt{n} \log n$ zeros in δ_n neighborhood of any point of the unit the circle. Thus the derived set, i.e., the set of limit points of the zeros of the polynomials appearing in the generalized Riesz product (R) is the full unit circle.

References

Chapter 1

[1] **H. Cramér.** *On The Structure of Purely Non-Deterministic Processes*, Arkiv för Matematik, **4, 2–3** (1961), 249–266.

[2] **P. R. Halmos.** *Introduction to Hilbert Space and The Theory of Spectral Multiplicity*, Second Edition, Chelsea Publishing Co. New York, 1957.

[3] **H. Helson.** *The Spectral Theorem*, Springer-Verlag Lecture Notes in Mathematics. No. 1227. (1986).

[4] **T. Hida.** *Canonical Representations of Gaussian Processes and Their Applications*, Mem. Coll. Sci. Kyoto, **A 33** (1960).

[5] **A. I. Plessner and V. A. Rokhlin.** *Spectral Theory of Linear Operators*, Uspekhi. Matem. Nauk, (N.S.) **1**(1946), 71–191.

[6] **M. H. Stone.** *Linear Transformations in Hilbert Space and Their Applications to Analysis*, A. M. S. Colloquium Publication, **vol 15** (1933).

Chapter 2

[1] **F. Riesz and Bela. Sz. Nagy.** *Functional Analysis*, Fredrick Ungar Publishing Co. New York, 1955.

Chapter 3

[1] **S. C. Bagchi, J. Mathew, and, M. G. Nadkarni.** *On Systems of Imprimitivity on Locally Compact Abelian Groups With Dense Actions*, Acta. Math., Uppsala, **133** (1974), 287–304.

[2] **M. Guenais.** *Une Majoration de la Multiplicité Spectrale d'opérateurs Associés à des Cocycles Réguliers*, Preprint (1997), University of Paris XIII.

[3] **H. Helson.** *Compact Groups With Ordered Duals*, Proc. London Math. Soc. **3** (1965), 14A 144–156.

[4] **A. Iwanik, M. Lemańczyk, D. Rudolph** *Absolutely Continuous Cocycles over Irrational Rotations*, Israel J. Math. **83** 1993 73–95.

[5] **A. B. Katok and A. M. Stepin.** *Approximation in Ergodic Theory*, Uspehi. Mat. Nauk.**22 No 5** (1967) 81–106, Russian Math. Surveys **22 No 5** (1967), 63–75.

[6] **G. W. Riley.** *On Spectral Properties of Skew Products over Irrational Rotations*, J. London Math. Soc. (2) **17** (1978), 161–164.

Chapter 4

[1] **R. V. Chacon.** *Approximation and Spectral Multiplicity, Contributions to Ergodic Theory and Probability*, Lecture Notes in Mathematics, No **160** (1970), Springer Verlag, 18–37.

[2] **A. del Junco.** *Transformations of Simple Spectrum which is not Rank One*, Canadian J. Math. **29** (1977), 655–663.

[3] **G. R. Goodson, J. Kwiatkowski, M. Lemanczyk, P. Liardet.** *On The Multiplicity Function of Ergodic Group Extensions of Rotations*, Studia Math. **102** (1992), 157–174.

[4] **A. Katok, A. Stepin.** *Approximations in Ergodic Theory*, Uspekhi. Mat. Nauk. **(22)** (1967), Russian Math. Surveys. **(22)** (1967), 77–102.

[5] **J. Kwiatkowski, M. Lemańczyk** *On the Multiplicity Function and Ergodic Group Extensions-II*, Studia Math. (1995)

[6] **E. A. Robinson, Jr.** *Transformations with Highly Non-Homogeneous Spectrum of Finite Multiplicity*, Israel J. Math. **56** no.1 (1986), 75–88.

[7] **E. A. Robinson, Jr.** *Non-Abelian Extensions have Non-simple Spectrum*, Compositio Math **65** (1988), 155–170.

Chapter 5

[1] **H. Anzai.** *Ergodic Skew Product Transformation on The Torus*, Osaka Math J. **3** (1951) 83–99.

[2] **E. A. Robinson, Jr.** *Non-Abelian Extensions Have Non-Simple Spectrum*, Composito Math. **65** (1988), 155–170.

[3] **K. Schmidt.** *Cocycles on Ergodic Transformation Groups*, MacMillan Co. of India, 1977.

Chapter 6

[1] **O. N. Ageev.** *Dynamical System With a Lebesgue Component of Even Multiplicity*, Mat. Sb **3** (7) (1988), 307–319 (in Russian).

[2] **A. Connes, J. Feldman, and B. Weiss.** *An Amenable Equivalence is Generated by a Single Transformation*, Ergodic Theory and Dynamical Systems, **1** (1981), 431–450.

[3] **H. Helson and W. Parry.** *Cocycles and Spectra*, Arkiv för Matematik, **16** (1978) 195–206.

[4] **M. Lemańczyk.** *Toeplitz Z_2-extensions,* Ann. Inst. H. Poincaré **24** (1988),1–43.

[5] **J. Mathew and M. G. Nadkarni.** *On Spectra of Unitary Groups Arising from Cocycles,* Arkiv for Matematik, **19** (1981), 229–237.

[6] **J. Mathew and M. G. Nadkarni.** *Measure Preserving Transformations Whose Spectrum Has Lebesgue Component of Multiplicity Two,* Bull. London Math. Soc. **16** (1984), 402–406.

[7] **M. Quéffelec.** Substitution Dynamical Systems-Spectral Analysis, Lecture Notes in Math. **1294**, Springer-Verlag 1988.

Chapter 7

[1] **J. Aaronson.** *The Eigenvalues of Non-Singular Transformations.* Israel J. Math. **45** (1983), 297–312.

[2] **J. R. Choksi and M. G. Nadkarni.** *Baire Category in Spaces of Measures, Unitary Operators, and Transformations.* Proc. Int. Conference on Invariant Subspaces and Allied Topics, (1986). Edited by H. Helson and B. S. Yadav, Narosa Publishers, New Delhi.

[3] **H. Furstenberg and B. Weiss.** *The Finite Multipliers of Infinite Ergodic Transformations.* Structures and Attractors in Dynamical Systems, Lecture Notes in Mathematics, **768** (1978), 128–132, Springer-Verlag, Berlin-Heidelberg-New York.

[4] **J. -P. Kahane and R. Salem.** *Ensembles Parfaits et Series Trignometriques.* Hermann, Paris, 1963.

[5] **K. Schmidt.** *Spectra of Ergodic Group Actions.* Israel J. Math. **41** (1982), 151–153.

[6] **K. Schmidt and P. Walters.** *Mildly Mixing Actions of Locally Compact Groups.* Proc. London Math. Soc. **(3)45** (1982), 506–508.

[7] **H. Weyl.** *Über die Gleichverteilung von Zahlen modulo Eins.* Selecta, Hermann Weyl, Birkhäuser, Basel, 1956.

Chapter 8

[1] **S. Banach.** *Théorie des Opérations Linéaires,* Chelsea, New York, 1963.

[2] **G. D. Birkhoff.** *Probability and Physical Systems,* Bull. Amer. Math. Soc. **138** (1932), 361–379. **Birkhoff:** *Collected Mathematical Papers,* vol. 2.

[3] **J. R. Choksi and S. Kakutani.** *Residuality of Ergodic Measurable Transformations and Transformations which Preserve an Infinite Measure,* Indiana University Mathematics Journal, **l28** (1979), 453–469.

[4] **J. R. Choksi and M. G. Nadkarni.** *Baire Category in Spaces of Measures, Unitary Operators and Transformations,* Proceedings of the International Conference on Invariant Subspaces and Allied Topics, University of Delhi, H. Helson and B. S. Yadav, Editors, Narosa Publishers, New Delhi.

[5] **I. P. Cornfeld, S. V. Fomin and Ya. G. Sinai.** *Ergodic Theory,* Springer-Verlag, New York, (1981).

[6] **A. del Junco.** *Disjointness of Measure Preserving Transformations, Minimal Self Joinings and Category,* Ergodic Theory and Dynamical Systems I, Progress in Mathematics 10, Birkhäuser, Boston, 1981, 81–89 .

[7] **H. Furstenberg.** *Disjointness in Ergodic Theory, Minimal Sets, and a Problem in Diophantine Approximation,* Math. Systems Theory, (1967) 1–50.

[8] **N. Friedman.** *Introduction to Ergodic Theory,* van Nostrand-Reinhold, New York, 1970.

[9] **F. Hahn and W. Parry.** *Some Characteristic Properties of Dynamical Systems with Quasi-Discrete Spectrum,* Math. Systems Theory, **2** (1968), 179–190.

[10] **P. R. Halmos.** *Lectures on Ergodic Theory,* Math. Soc. Japan Publication, Tokyo 1956. Reprinted Chelsea New York, 1960.

[11] **P. R. Halmos.** *Approximation Theories for Measure Preserving Transformations,* Trans. Amer. Math. Soc. **55** (1944) 1–18.

[12] **P. R. Halmos.** *In General a Measure Preserving Transformation is mixing,* Ann. of Math. **45** (1944), 786–792.

[13] **J. M. Hawkins and E. A. Robinson Jr.** *Approximately Transitive Flows and Transformations Have Simple Spectrum,* Preprint 1985.

[14] **A. Katok and A. M. Stepin.** *Approximation in Ergodic Theory,* Russian. Math. Surveys, **22** (1967), 77–102.

[15] **A. Katok.** *Approximation and Genericity in Abstract Ergodic Theory,* Notes 1985.

[16] **J. C. Oxtoby and S. Ulam.** *Measure Preserving Homeomorphisms and Metric Transitivity,* Ann. Math. (2) **42** (1941), 874–920.

[17] **K. Petersen.** *Ergodic Theory,* Cambridge Studies in Advanced Mathematics; 2, Cambridge University Press, 1983.

[18] **V. A. Rokhlin.** *New Progress in the Theory of Transformations with Invariant Measure,* Russian. Math. Surveys, **15** (1960), 1–22.

[19] **V. A. Rokhlin.** *The General Measure Preserving Transformation is Mixing,* Dokl. Acad. Sci. USSR, **3** (1948), 349–358.

[20] **H. L. Royden.** Real Analysis, Edition 3, MacMillan Publishing Co., (1989), New York.

[21] **B. Simon.** *Operators with Singular Continuous Spectrum: I. General Operators,* Annals of Math. **141** (1995), 131–145.

[22] **A. M. Stepin.** *Spectral Properties of Generic Dynamical Systems,* Math. USSR Izvestiya, **29** (1987), No.1.

Chapter 9

[1] **P. Halmos.** *Measure Theory,* D. Van Nostrand Company, New York, 1950.

[2] **B. Host.** *Mixing of All Orders and Pairwise Independent Joinings of Systems with Singular Spectrum,* Israel Journal of Mathematics, **76** (1991), 289–298.

[3] **B. Host, J. F. Méla, F. Parreau.** *Non-Singular Transformations and Spectral Theory,* Bull. Soc. Math. France. **119** (1991), 33–90.

[4] **B. Host, F. Parreau.** *The Generalised Purity Law for Ergodic Measures: A Simple Proof,* Colloquium Mathematicum, **Vol LX/LXI** (1990), 206–212.

[5] **G. W. Mackey.** *Borel Structure in Groups and Their Duals,* Trans. Amer. Math. Soc. **85** (1957), 134–185.

[6] **V. Mandrekar and M. Nadkarni.** *On Ergodic Quasi-invariant Measures on The Circle Group,* J. Funct. Anal. **3** (1969), 157–163.

[7] **W. Rudin.** *Fourier Analysis on Groups,* Interscience Tracts in Math. **12**, Wiley, New York,1967.

[8] **A. Weil.** *L'Intégration dans les groupes Topologiques et ses Applications,* Paris, 1940.

Chapter 10

[1] **I. Assani.** *Multiple Recurrence and Almost Sure Convergence for Weakly Mixing Dynamical Systems,* Preprint, University of North Carolina, Chapel Hill, to appear in Israel Journal of Mathematics.

[2] **B. Host.** *Mixing of All Orders and Pairwise Independent Joinings of Systems With Singular Spectrum,* Israel. Jour. Math.**76** (1991), 289–298.

[3] **S. Kalikow.** *Two Fold Mixing Implies Three Fold Mixing for Rank One Transformations,* Ergodic Theory and Dynamical Systems. **4** (1984), 237–259.

Chapter 11

[1] **J. Aaronson and M. Nadkarni.** L^∞ *eigenvalues and* L^2 *spectra of Non-Singular Transformations,* Proc. London Math. Soc. (3) **55** (1988), 538–570.

[2] **S. Banach.** *Théorie des Opérations Linéaires,* Chelsea, New York, 1963.

[3] **B. Host, J. F. Méla and F. Parreau.** *Non-Singular Transformations and Spectral Analysis of Measures,* Bull. Soc. Math. France, **119** (1991), 33–90.

[4] **C. C. Moore and K. Schmidt.** *Coboundaries and Homomorphisms for Non-Singular Group Actions and a Problem of H. Helson,* Proc. London. Math. Soc. (3) **40** (1980), 443–475.

[5] **K. Schmidt.** *Spectra of Ergodic Group Actions,* Israel J. Math. **41** (1982), 151–153.

Chapter 12

[1] **H. Helson.** *Analyticity on Compact Abelian Groups,* Algebras in Analysis, editor J. Williamson, Academic Press, New York, 1975 1–62.

[2] **H. Helson and D. Lowdenslager.** *Invariant Subspaces*, Proc. Int. Symp. on Linear Spaces, Jerusalem (1961), 251–262.

[3] **G. W. Mackey.** *A Theorem of Stone and von Neumann*, Duke Math J. **16** (1949), 313–326.

[4] **G. W. Mackey.** *Infinite Dimensional Group Representations*, Bull. Amer. Math. Soc., **69** (1963), 628–686.

[5] **V. S. Varadarajan.** *Geometry of Quantum Theory*, **II** van Nostrand-Reinhold, New York, 1970.

Chapter 13

[1] **J. Aaronson.** *The Intrinsic Normalising Constants of Transformations Preserving Infinite Measure*, J. Analyse Math. **49** (1987) 239–270.

[2] **J. Aaronson, M. Nadkarni.** L_∞ *eigenvalues and* L_2 *spectra of Non-Singular Transformations*, Proc. London. Math. Soc. (3) **55** (1987), 538–570.

[3] **S. Bagchi, J. Mathew, M. Nadkarni.** *On Systems of Imprimitivity on Locally Compact Groups With Dense Actions*, Acta. Math. **133** (1974), 287–304.

[4] **H. Helson.** *Cocycles On The Circle*, J. Operator Theory **16** (1986), 189–199.

Chapter 14

[1] **B. Host, J. F. Méla and F. Parreau.** *Non-Singular Transformations and Spectral Analysis of Measures*, Bull. Soc. Math. France. **119** (1991), 33–90.

[2] **J. F. Méla.** *Groupes de Valeurs Propres des Systèmes Dynamiques et Sous-groupes Saturés du Circle*, C.R. Acad. Sci. Paris, Série I Math. **296** (1983), 419–422.

[3] **C. C. Moore and K. Schmidt.** *Coboundaries and Homomorphisms for Non-Singular Group Actions and a Problem of H. Helson*, Proc. London Math. Soc. (3) **40** (1980), 443–475.

[4] **K. Schmidt.** *Spectra of Ergodic Group Actions*, Israel J. Math. **41** (1982), 151–153.

Chapter 15

[1] **T. Adams.** *Classical Staircase Construction is Mixing*, Preprint

[2] **J. R. Baxter.** *A Class of Ergodic Automorphisms*, Ph.D. thesis, Univ. of Toronto, 1969.

[3] **E. Beller.** *Polynomial Extremal Problems in* L^p, Proc. Amer. Math. Soc.**30** (1971), 250–259.

[4] **G. Brown and A. H. Dooley.** *Odometer Actions on G-measures*, Ergodic Theory and Dynamical Systems,**11** (1991) 297–307.

[5] **G. Brown and A. H. Dooley.** *Dichotomy Theorems for G-measures*, To appear in the International Journal of Mathematics.

[6] **J. Bourgain.** *On the Spectral Type of Ornstein's Class One Transformations*, Israel J. of Math. **84** (1993), 250–259.

[7] **R. V. Chacon.** *A Geometric Construction of Measure Preserving Transformations*, Proc. Fifth. Berkeley Symposium on Mathematical Statistics and Probability, Berkeley and Los Angeles, University of California Press **vol 2** part 2 (1965), 335–360.

[8] **J. Choksi and M. Nadkarni.** *Maximal Spectral Type of a Rank One Transformation*, Canad Math Bull **37** (1), (1994), 29–36.

[9] **J. Choksi and M. Nadkarni.** *The Group of Eigenvalues of Rank One Transformation*, Canad Math Bull **37** (1), (1994), 29–36.

[10] **A. del Junco, M. Rahe and L. Swanson.** *Chacon's Automorphism Has Minimal Self Joinings*, Journal D'Analyse Mathématique, **37** (1980), 276–284.

[11] **J. L. Doob.** *Stochastic Processes* Wiley Interstice, New York, 1953.

[12] **N. Friedman.** *Replication and Stacking in Ergodic Theory,* Amer. Math. Monthly, **99** (1992), 31–41.

[13] **Mélanie Guenais** *Morse Cocycles and Simple Lebesgue Spectrum*, preprint, University of Paris XIII.

[14] **B. Host, J.-F. Méla, F. Parreau.** *Non-Singular Transformations and Spectral Analysis of Measures*, Bull. Soc. Math. France **119** (1991), 33–90.

[15] **El Abdalaoui El Houcein.** *La Singularité Mutuelle Presque Sûre Du Spectre Des Transformations D'Ornstein*, Preprint 1997, University of Rouen, Rouen, France.

[16] **Y. Ito, T. Kamae and I. Shiokawa.** *Point Spectrum and Hausdorff Dimension*, Number Theory and Combinatorics, edited by J. Akiyama et al, World Scientific Publishing Co. Tokyo,(1985), 209–277.

[17] **J.-P. Kahane.** *Sur les Polynômes à Coefficients Unimodulaires*, Bull. London. Math. Soc. **12** (1980), 321–342.

[18] **I. Klemes.** *The Spectral Type of the Staircase Transformation*, Tohoku. Math. Journal, **48** (1996), 247–258.

[19] **I. Klemes and K. Reinhold.** *Rank One Transformations With Singular Spectral Type*, Israel. Jour. Math. **98**,(1997), 1–14.

[20] **F. Ledrappier.** *Des Produits de Riesz comme Measure Spectrales*, Ann. Inst. Henri Poincaré **6**(4) (1970), 335–344.

[21] **D. J. Newman.** *An Extremal Problem for Polynomials*, Proc. Amer. Math. Soc. **16** (1965) 1287–1290.

[22] **D. S. Ornstein.** *On The Root Problem In Ergodic Theory*, Proc. Sixth Berkeley Symposium on Mathematical Statistics and Probability, Berkeley and Los Angeles, University of California Press **vol 2** (1970), 348–356.

[23] **M. Osikawa.** *Point Spectrum of Non-Singular Flows*, Publ. Res. Inst. Math. Sci. Kyoto. Univ. **13** (1977), 167–172.

[24] **J. Peyriére.** *Étude de Quelques Propriétés des Produits de Riesz*, Ann. Inst. Fourier (2) **25** (1975),127–169.

[25] **F. Riesz.** *Über die Fourierkoeffizienten einer stetigen Funktion von be-schränkter Schwankung*, M.Z. **2** (1918), 312–315.

[26] **A. Zygmund.** *Trigonometric Series*, Second Edition, Cambridge University Press, 1968, 208–212.

Chapter 16

[1] **G. Goodson and M. Lemańczyk.** *Transformations Conjugate to Their Inverses Have Even Essential Values*, Proc. Amer. Math. Soc. **124** (1996) 2703–2710.

[2] **N. A. Friedman.** *Replication and Stacking in Ergodic Theory*, Amer. Math. Monthly, **99** (1992), 31–44.

[3] **N. A. Friedman and D. S. Ornstein.** *On Partially Mixing Transformations*, Indiana Univ. Math. J. **20** (1971), 767–775.

[4] **G. R. Goodson, A. del Junco, M. Lemańczyk, D. J. Rudolph.** *Ergodic Transformations Conjugate to Their Inverses by Involutions*, Ergodic Theory and Dynamical Systems **16** (1996), 97–124.

[5] **M. Lemańczyk.** *Introduction to Ergodic Theory from the Point of View of Spectral Theory*, Lecture Notes on the Tenth Kaisk Mathematics Workshop, Geon Ho Choe (ed), Korea Advanced Institute of Science and Technology, Math. Res. Center, Taejon, Korea.

[6] **D. S. Ornstein.** *On the Root Problem in Ergodic Theory*, Proc. Sixth Berkeley Symposium on Mathematical Statistics and Probability, Berkeley and Los Angeles, University of California Press, **vol 2** 1970 348–356.

Chapter 17

[1] **El. H. El Abdalaoui.** *A new class of rank-one transformations with singular spectrum*, Ergodic Theory Dynam. Systems, **27** (2007), no. 5, 1541-1555.

[2] **El. H. El Abdalaoui.** *Ergodic Banach problem on simple Lebesgue spectrum and flat polynomials* Arxiv 1508.06439v.

[3] **El. H. El Abdalaoui.** *On Erdos flat polynomials problem, Chowla conjecture and Riemann hypothesis* Arxiv 1609.03435v

[4] **El. H. El Abdalaoui, M. G. Nadkarni** *Calculus of Generalized Riesz Products*, Contemporary Mathematics, **631**(2015), 143-178.

[5] **El. H. El Abdalaoui, M. G. Nadkarni** *A non-singular transformation whose spectrum has Lebesgue component of Multiplicity one*, Ergodic Theory and Dynamical Systems, **36**(2016) 671-681.

[6] **P. Borwein, M. J. Mossinghoff.** *Barker sequences and flat polynomials*, Number Theory and Polynomials, 71-88, Lond. Math. Soc. Lecture Notes Series, 352, Cambridge Univ Press, Cambridge, (2008).

[7] **J. Bourgain.** *On the spectral type of Ornstein class one transformations*, Israel J. Math., **84** (1993), 53-63.

[8] **P. Brown, T. Erdélyi, F. Littmann.** *Polynomials with coefficients from a finite set,* Trans. Amer. Math. Soc. **360**(10), 5145-5154, (2008).

[9] **G.Brown, W.Moran.** *On orthogonality of Riesz products,* Proc. Cambridge Philos. Soc., **76** (1974), 173-181.

[10] **T. Downarowich, Y. Lacroix.** *Merit Factors and Morse Sequences* Theoretical Computer Science, **(209)** (1998), 377-387.

[11] **M. Guenais.** *Morse cocycles and simple Lebesgue spectrum* Ergodic Theory Dynam. Systems,**19** (1999), no. 2, 437-446.

[12] **K. Hoffman.** *Banach spaces of analytic functions,* Reprint of the 1962 original. Dover Publications, Inc., New York, 1988.

[13] **I .Klemes, K. Reinhold.** *Rank one transformations with singular spectre type,* Isr. J. Math., **98** (1997), 1-14.

[14] **A. M. Odlyzko, B. Poonen.** *Zeros of polynomials with 0,1 coefficients,* l'Enseign. Math., **39** (1993), pp. 317-348.

[15] **A. A. Prikhodko.** Flat trigonometric sums and ergodic flows with simple spectrum. Arxiv 1003. 2808v

[16] **S. Ulam.** *Problems in Modern Mathematics, p 76,* Science Edition, John Wiley, New York, 1964.

Index

Texts and Readings in Mathematics